34th PIANC World Congress 2018

Connecting Maritime Hubs Globally

Dredging

Panama City, Panama
7 - 11 May 2018

ISBN: 978-1-5108-8140-2

Printed from e-media with permission by:

Curran Associates, Inc.
57 Morehouse Lane
Red Hook, NY 12571

Some format issues inherent in the e-media version may also appear in this print version.

Printed by Curran Associates, Inc. (2019)

For permission requests, please contact
The World Association for Waterborne Transport Infrastructure (PIANC)
at the address below.

The World Association for Waterborne Transport Infrastructure (PIANC)
General Secretariat
Boulevard Roi Albert II, 20 – B. 3
B-1000 Brussels
Belgium

Phone: +32 2 553 71 61

info@pianc.org

Additional copies of this publication are available from:

Curran Associates, Inc.
57 Morehouse Lane
Red Hook, NY 12571 USA
Phone: 845-758-0400
Fax: 845-758-2633
Email: curran@proceedings.com
Web: www.proceedings.com

TABLE OF CONTENTS

COMMITTEES

INTERNATIONAL ORGANIZING COMMITTEE (IOC)

LOCAL ORGANIZING COMMITTEE (LOC)

Mr. Manuel Benítez, Congress Co-Chair (ACP)
Mr. Rogelio Gordón, Congress Co-Chair (ACP)
Mr. Oscar Vallarino, Local Executive Committee (ACP)
Ms. María Teresa Maure (ACP)
Mr. Ricardo Álvarez (ACP)

INTERNATIONAL SCIENTIFIC COMMITTEE (ISC)

Dr. Philippe Rigo, Vice President of the SC (Belgium)
Dr. Michael Fastenbauer, Chair (Austria)
Mr. Jean Michel Hiver (Belgium)
Mr. Jose Sanchez (USA)
Mr. Wu Peng, Chair (China)
Mr. Jim Stirling, Chair (UK)
Mr. Fabrice Daly, Chair (France)
Mr. Rik Thomas (Belgium)
Mr. Juergen Troegl, Chair (Austria)
Mr. Leonel Temer (Argentina)
Ms. Milou Wolters, Chair (Netherlands)
Mr. Stefan Devocht (Belgium)
Dr. Carsten Thorenz, Chair (Germany)
Dr. Tom De Mulder, Chair (Belgium)
Mr. Holger Schüttrumpf (Germany)
Mr. John Clarkson, Chair (USA)
Mr. Timothy Paulus, Chair (USA)
Dr. Jasna Muskatirovic (Serbia)
Mr. Ian White, Chair (UK)
Mr. Francisco Esteban Lefler, Chair (Spain)
Mr. Ignacio Sanchidrian Vidal, Chair (Spain)
Mr. Hartmut Bruehl, Chair (Germany)

Mr. H.N. Aswath (India)
Mr. Craig Philip, Chair (USA)
Mr. Edward Schmeltz (USA)
Mr. Hans Burcharth (Denmark)
Dr. Mohammadreza Allahyar (Iran)
Ms. Katrina Dodd (Australia)
Mr. Esteban Biondi, Chair (USA)
Mr. Pablo Arecco, Chair (Argentina)
Mr. Nicholas Pansic, Chair (USA)
Dr. Todd Bridges (USA)

LOCAL SCIENTIFIC COMMITTEE (LSC)

Dr. Dimas E. Portillo L., President of the SC (ACP)
Mr. Abdiel Perez (ACP)
Mr. Rigoberto Delgado (ACP)
Mr. Antonio Abrego (ACP)
Mr. Ricardo Álvarez (ACP)
Dr. Ricardo Ungo (ACP)
Mr. Juan Wong, Chair (Panama)

COPYRIGHT AND REPRODUCTION

PREFACE

This book of abstracts contains 284 abstracts presented at the 34th PIANC World Congress held in Panama, Republic of Panama, from 7 to 11 May 2018. PIANC World Congress presents and discusses topics relevant to the waterborne transport and infrastructure sector. It rotates among PIANC's member countries every four years and it is open to members and non-members. Participants, have the opportunity to exchange knowledge and experiences.

PIANC's goal is to advance the sustainable development of shallow and deep-draft navigation issues including dredging and dredged material disposal, navigation and pot infrastructure, recreational navigation and related environmental matters.

The Book of Abstract is arranged according to the main Themes of the Congress namely: Inland Navigation, Dredging, Logistics & Infrastructure, Ports, Marinas, and Environment. Inside each Theme, the abstracts are arranged by the topics of the Theme.

Inland Navigation

- Waterway infrastructures: locks, weirs, riverbanks, others.
- Advances in navigation locks design after the Panama Canal Expansion
- Inland navigation channels: safety and reliability
- Integration of inland waterways into inter-modal supply chain
- Inland navigation, waterways, ports & terminals
- Systems and Infrastructures: design & management for inland navigation projects
- River Information Services (RIS, AIS, others)
- River navigation flow control

- Salt water intrusion
- Maintenance and operation of IW, transport & infrastructures
- Inland navigation in South America

Dredging (in the framework of port and navigation projects)

- Effective planning and execution of dredging projects
- Lessons learned from dredging projects worldwide
- Risk management in dredging projects
- Disposal site planning taking into account alternative land uses
- Current dredging & management innovations

Logistics & Infrastructure

- Challenges for capitalizing navigational channels bordering lands
- Synergies among airports, ports and inter-modal assets
- Integrated management of global supply chains
- Integrating ports and economic special Zones
- Provision of added value logistic services in maritime hubs
- Project management in the Panama Canal Expansion

Ports

- Impact and innovative solutions for hinterland ports
- Dry ports: better and more efficient
- Ports of the future: technologies, automation, traceability
- Port management models
- Maritime Port planning and operations
- Inter-modal connections
- Coastal and Port Engineering (in relation with navigation)
- Storm Surge & Tsunami barriers and Flood protection gates

Marinas

- Sustainable and resilient marina design-Marinas Working with Nature (best practices and case studies)
- Marinas as part of Tourism, Ports and Urban Master Plans (best practices and case studies)
- Planning Framework for Recreational & Tourism Navigation Infrastructure Systems (goals, strategies and challenges)
- Regulatory Framework and Private Marina Development in Countries in Transition (best practices and case studies)
- Regional Cooperation and Standardization of Recreational Navigation Information Systems Transition (best practices and case studies)
- Recreational components of mitigation plans for large navigation infrastructure projects

Environment

- Environmental management in navigation
- Climate change and emissions, energy efficiency, international regulations, carbon markets
- Fresh water availability for operations
- New technologies on infrastructure, pollution prevention, port reception facilities and ballast water
- Multiple purpose water resource systems (transport, energy, recreation, ecosystems, watersheds, potable water, ...)
- Navigation as a green transportation mode
- Societal awareness and responsibility, combining economic growth, environment / sustainability and welfare
- Climate Change Mitigation in the Maritime Industry

All these abstracts represent an invaluable source of information, knowledge, experience, and ideas useful for students, engineers, researchers, scientists, authorities and for everyone interested in the fields covered in this World Congress.

Name___________________________________
Title___________________________________
Position_________________________________
Affiliation______________________________

LONG-TERM SEDIMENT MANAGEMENT PLANNING AT NORTH AMERICA'S LARGEST PORT COMPLEX: BALANCING THE NEED TO ACCOMMODATE THE LARGEST SHIPS WHILE COMPLYING WITH COMPLEX ENVIRONMENTAL REQUIREMENTS

by

Kathryn Curtis[1], Matthew Arms[2], Shelly Anghera, Ph.D.[3], and Steve Cappellino[4]

ABSTRACT

Port growth is an important economic investment and ports are performing infrastructure improvements that include larger cranes, higher capacity backlands for quicker through port transfer, greater inland infrastructure, and deeper drafts. The dimensions of the world-wide fleet of container vessels have increased significantly in the last ten years, and it is anticipated that this trend will continue into the future as shipping companies continue to consolidate. As a result, ports throughout the United States have been actively positioning themselves to accommodate larger vessels to maintain or grow trade opportunities. While market pressures are driving the Ports of Long Beach and Los Angeles to deepen/redevelop their berths, tightening regulations that limit cost effective sediment management options like ocean disposal have created new challenges for Port staff.

Planning for large dredging/redevelopment projects frequently entails unique engineering and environmental challenges. The financial and environmental feasibility of these projects is often dependent on the management of contaminated sediments or large quantities of clean sediment that must be undertaken within region-specific regulatory requirements. In California, several regulatory authorities oversee the movement and disposal of sediment. They make up the Los Angeles Regional Contaminated Sediments Task Force and include: U.S. Army Corps of Engineers, U.S. Environmental Protection Agency, National Marine Fisheries Service, U. S. Fish and Wildlife Service, California Coastal Commission, Los Angeles Regional Water Quality Control Board, and California Department of Fish and Wildlife. Dredging and disposal activities are also watched closely by local non-governmental organizations to ensure that regional initiatives like the promotion of beneficial reuse opportunities are being met. In the Los Angeles region, the preferred management strategy for clean sediments is beach nourishment and port fills. Management strategies for contaminated sediments, in order of preference, include beneficial reuse of sediments in construction fill (e.g., nearshore confined disposal facility), temporary storage in an approved upland area (until a fill project becomes available), treatment and reuse as a marketable product (e.g., cement-stabilized fill), or for other beneficial upland placement

[1] Marine Environmental Supervisor, Environmental Management Division, Port of Los Angeles, USA. kcurtis@portla.org

[2] Assistant Director of Environmental Planning, Port of Long Beach, USA. matthew.arms@polb.com

[3] Principal Scientist, Latitude Environmental, Inc., USA. shelly@latitudeenv.com

[4] Principal Scientist, Anchor QEA, LLC, USA. scappellino@anchorqea.com.

areas. The ports do not necessarily have planned fill opportunities to align with maintenance and capital dredging projects; therefore, the ports have been helping the regulators recognize the need for regional opportunities to manage both clean and contaminated sediments effectively and efficiently. Both ports have developed sediment management guidance documents and contaminated sediment management plans to provide clarity in the ports' decision processes, and to prioritize management strategies that are feasible for implementation within a working port. Through frequent communication with the regulators, the ports have been able to generate renewed interest in confined aquatic disposal facilities and shallow water habitat enhancement opportunities because of the need to identify economical and logistically feasible management alternatives for clean and contaminated sediments beyond use as port fill material. Creation of biologically valuable habitat with dredged material has the added value of providing mitigation credit for port development projects.

This talk focuses on the Ports' sediment management challenges and discusses the research conducted over the last few years to identify suitable sediment management alternatives at the nation's largest ports, Port of Los Angeles and Port of Long Beach.

34th PIANC WORLD CONGRESS PANAMA 2018 May 7 to 11

CANAL DE PANAMÁ

Long-term Sediment Management Planning at North America's Largest Port Complex

Kathryn Curtis, Port of Los Angeles
Matthew Arms, Port of Long Beach
Shelly Anghera, Latitude Environmental, Inc.
Steve Cappellino, Anchor QEA

THE PORT OF LOS ANGELES · The Port of LONG BEACH · LATITUDE ENVIRONMENTAL · ANCHOR QEA

34th PIANC WORLD CONGRESS PANAMA 2018 May 7 to 11

Ports of Los Angeles and Long Beach

34th PIANC WORLD CONGRESS PANAMA 2018 May 7 to 11

CANAL DE PANAMÁ

Global Trade Leaders

- Container, breakbulk, liquid bulk, dry bulk terminals
- Connectivity to all US markets and gateway to Pacific Rim
- 2017 container volumes
  - POLA – 9.3 TEUs
  - POLB -

34th PIANC WORLD CONGRESS PANAMA 2018 May 7 to 11

CANAL DE PANAMÁ

Deepening Navigation Depths

- Deepen/maintain channels and berths for increasingly larger ships
- Funding/coordination with US Army Corps of Engineers on deepening federal channels
- Balance need to accommodate the largest ships while complying with complex environmental requirements

34th PIANC WORLD CONGRESS
PANAMA 2018 May 7 to 11
CANAL DE PANAMÁ

Dredged Material Management

- Strong environmental regulations/standards
- Contaminated Sediments Task Force – multiple regulatory agencies oversee dredge projects
- Emphasis on beneficial reuse of dredged material
- Valuable biological resources in harbor – mitigation requirements for fills
- Challenges with dredged material disposal

34th PIANC WORLD CONGRESS
PANAMA 2018 May 7 to 11
CANAL DE PANAMÁ

Port-Specific Projects Overview

- Each port has its own approach, resources, and obstacles
- Sediment management challenges
- Environmental challenges
- Engineering challenges
- Long-term planning
- Lessons learned

34th PIANC WORLD CONGRESS PANAMA 2018 May 7 to 11

CANAL DE PANAMÁ

Port of Los Angeles

34th PIANC WORLD CONGRESS PANAMA 2018 May 7 to 11

Port of Los Angeles Channel Deepening Project

- 2002 - 2013
- -53 feet depth
- 15,000,000 cy dredged
- 100% beneficial use
- $370 million
- Cost sharing with ACOE from WRDA 2000

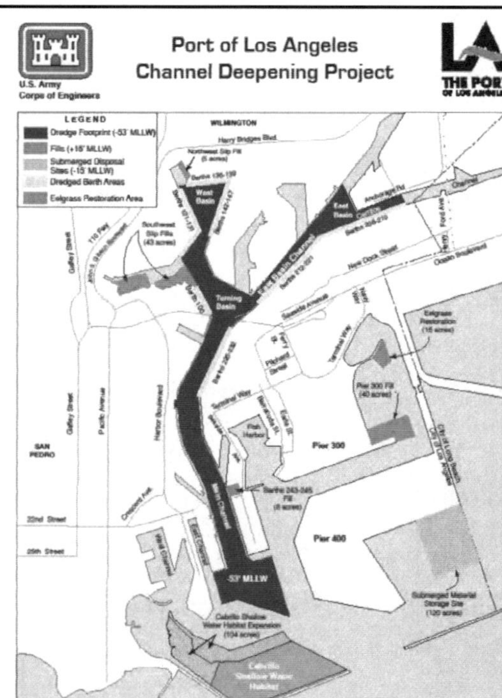

34th PIANC WORLD CONGRESS PANAMA 2018 May 7 to 11

CANAL DE PANAMÁ

Sediment Management Challenges

- Issue with native material not going to ocean disposal site due to naturally occurring metals
- Poor quality material necessitated mining sand for fills
- Project delays
  - Permits
  - Legal challenge
  - Remediation dredging project added
  - Additional in-harbor disposal capacity needed

34th PIANC WORLD CONGRESS PANAMA 2018 May 7 to 11

CANAL DE PANAMÁ

Sediment Management Solutions

- Habitat improvement
  - 104 acres of Shallow Water Habitat
  - 15-ac Eelgrass Restoration Site
- 96 acres of new land fills, including two Confined Disposal Facilities
- Surcharge for new fill – eventually disposed in Shallow Water Habitat
- Creation of 200 acre in-harbor Submerged Material Storage Site

34th PIANC WORLD CONGRESS PANAMA 2018 May 7 to 11

CANAL DE PANAMÁ

Environmental Challenges

- Regulatory
  - COE was lead on project
  - Port/COE coordination with other agencies early and throughout project
- Biological mitigation requirements
  - Created Shallow Water Habit and Eelgrass Restoration Site to partially compensate for impacts from new fills
  - Also used banked mitigation credits

34th PIANC WORLD CONGRESS PANAMA 2018 May 7 to 11

CANAL DE PANAMÁ

Engineering Challenges

- Removing surcharge and building a conveyor bridge over a working terminal
- Maintaining line and grade with bottom dumps at shallow water habitat area
- Equipment and key personnel availability over a 10-year project

34th PIANC WORLD CONGRESS PANAMA 2018 May 7 to 11

Lessons Learned

- Maximizing dredged material disposal in-harbor allows for beneficial creation of new land fills for Port uses and enhanced biological habitat, and also reduces disposal costs
- Shorter projects are better than longer projects

34th PIANC WORLD CONGRESS PANAMA 2018 May 7 to 11

CANAL DE PANAMÁ

Port of Long Beach

34th PIANC WORLD CONGRESS PANAMA 2018 May 7 to 11

CANAL DE PANAMÁ

Deepening Project

- 3-3-3 USACE process
- 8 million cy
- -55' + 2 ft overdepth
- -73' approach

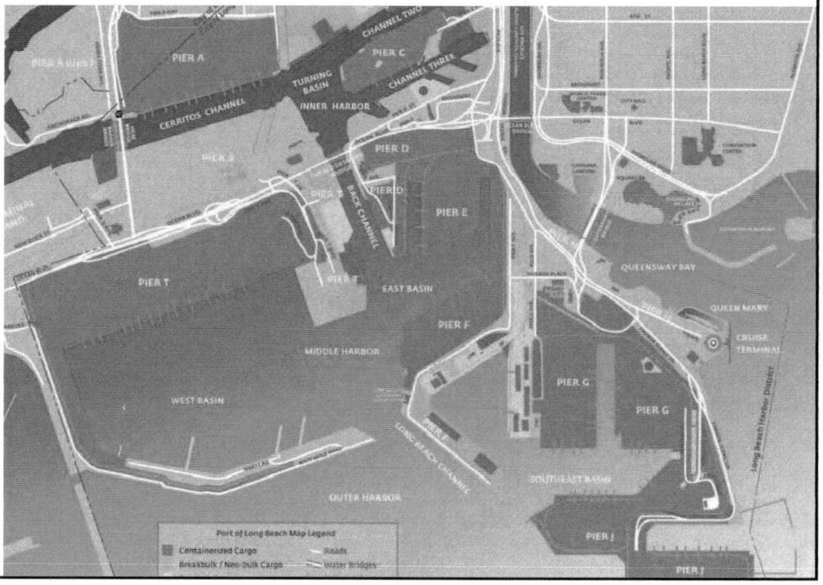

34th PIANC WORLD CONGRESS PANAMA 2018 May 7 to 11

CANAL DE PANAMÁ

Solution: Long-Term Management Planning

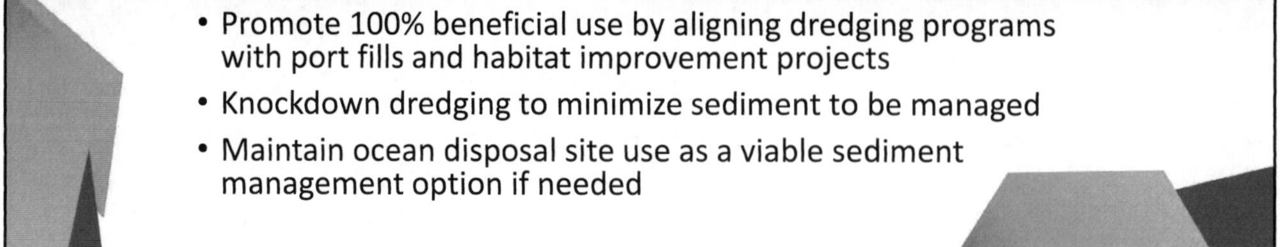

- Promote 100% beneficial use by aligning dredging programs with port fills and habitat improvement projects
- Knockdown dredging to minimize sediment to be managed
- Maintain ocean disposal site use as a viable sediment management option if needed

34th PIANC WORLD CONGRESS PANAMA 2018 May 7 to 11

New Management Alternatives to Be Explored

- Conversion of temporary aquatic sediment storage sites to confined aquatic disposal sites
- Evaluation of feasibility of new shallow water habitat enhancement areas
- Evaluation of new temporary sediment storage site areas

34th PIANC WORLD CONGRESS PANAMA 2018 May 7 to 11

CANAL DE PANAMÁ

Outer Harbor Sediment Placement Site

- South area = 1.4 million cy capacity for clean material
- North area = 1.3 million cy capacity for non-open ocean suitable material

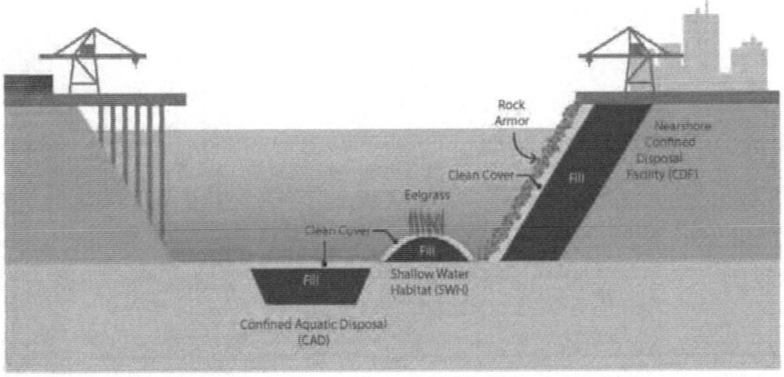

34th PIANC
WORLD CONGRESS
PANAMA 2018 May 7 to 11

New Engineering Challenges

- Electric dredges for air quality
- Reduced competition on west coast for dredge contractors
- New limits for post-dredge surface concentrations

34th PIANC
WORLD CONGRESS
PANAMA 2018 May 7 to 11

CANAL DE PANAMÁ

Lessons Learned

- Develop as many beneficial use opportunities as possible
- Work to provide flexibility of sediment management options in permits
- Provide opportunities for region to benefit from sediment management strategy

34th PIANC
WORLD CONGRESS
PANAMA 2018 May 7 to 11

CANAL DE PANAMÁ

- Kathryn Curtis: kcurtis@portla.org
- Matthew Arms: matthew.arms@polb.com
- Shelly Anghera: shelly@latitudeenv.com
- Steve Cappellino: scappellino@anchorqea.com

PIAÇAGUERA CHANNEL DREDGING CASE: CONFINED AQUATIC DISPOSAL - CAD AS AN ALTERNATIVE FOR THE DESTINATION OF SEDIMENTS NOT AVAILABLE TO THE OCEAN DISPOSAL.

by

Mauricio Torronteguy[1], Juliana M. Menegucci[2], Anderson Saraiva[3], Flavia Camara[4], Nathália Castro[5], André Marques[6] and Laiza Coelho[7]

ABSTRACT

This article presents the history, criteria and premises that support decision making for the use of an underwater Confined Aquatic Disposal - CAD in the interior of the Santos estuary, São Paulo, Brazil, as an alternative of not suitable for oceanic disposal dredging material of the Piaçaguera Channel, as well as the constructive methodology, technical management, environmental controls and learned lessons in the distinct phases of the CAD implementation.

1. INTRODUCTION

The Piaçaguera Channel, located in the Santos Estuary, São Paulo, Brazil, is an important navigation route linking the Terminal Integrador Portuário Luiz Antônio Mesquita - TIPLAM and USIMINAS terminal to the navigation channel of the Port of Santos, considered the most important port of South America (Figure 1).

For years the Piaçaguera Channel could not be fully dredged due to the existence of environmental restrictions on direct oceanic disposal related to the presence of inadequate quality sediments in the channel.

During 2015 and 2016 the environmental restoration process and reestablishing of navigations conditions were initiated with the dredging of approximately 300,000.00 cubic meters of sediment not suitable for oceanic disposal, while TIPLAM new berth implementation dredging occurred. The initial solution for this issue was the use of geobags for sediment confinement at a CDF - Confined Disposal Facility, built at a site close to the TIPLAM area.

As defined by the IADC (2010) a CDF is an area specifically designed to contain contaminated dredged material where it is possible to control releases of lower quality material to the environment occurring during the discharge of treated effluents, so, free of contamination. Dykes or other structures such as geobags can be used to insulate dredged material and the main purpose of a CDF is to keep dredged material solids in the confined area, and to treat the process water so that stay free of contaminants and be discarded or reused.

Due to the significantly larger volume of material that needed to be dredged for maintenance and deepening of the Piaçaguera Channel, in the order of 2.3 million cubic meters, it was necessary to find an alternative solution from the one used in the TIPLAM berths implementation.

[1] Oceanographer, Msc., Atlântico Sul Consultoria, Brazil, mauricio@atlsul.com.br
[2] Civil engineer and oceanographer, Atlântico Sul Consultoria, Brazil, juliana@atlsul.com.br
[3] Mechanical enginner, VLI Logística, Brazil, anderson.saraiva@vli-logistica.com.br
[4] Chemical enginner, VLI Logistica, Brazil, flavia.camara@vli-logistica.com.br
[5] Biologist, VLI Logistica, Brazil, nathalia.castro@vli-logistica.com.br
[6] Civil engineer, Atlântico Sul Consultoria, Brazil, andre@atlsul.com.br
[7] Electrical enginner, VLI Logistica, Brazil, laiza.coelho@vli-logistica.com.br

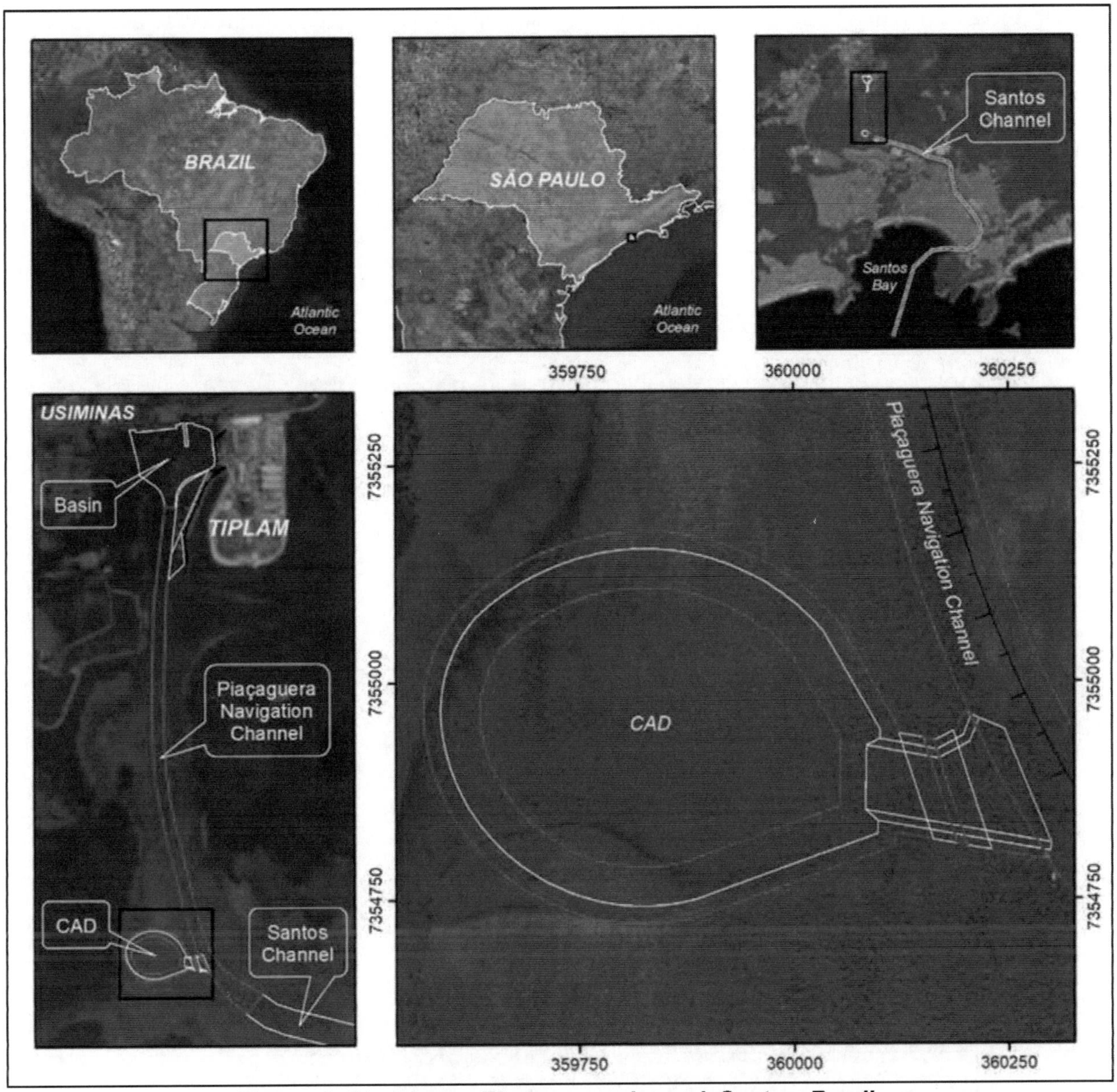

Figure 1: Study area. Piaçaguera channel, Santos, Brazil.

In this way, several engineering and environmental studies were carried out in the search for a technical, economic and environmental solution that would allow the safe handling and disposal of the sediments not suitable for oceanic disposal from the Piaçaguera Channel. The decision matrix indicated the better alternative as an underwater Confined Aquatic Disposal - CAD cell in the interior of the Santos estuary, near the Piaçaguera Channel, in an area duly licensed and authorized by the Brazilian authorities based on rigorous and complex environmental studies carried out to obtain the permitting.

The CAD's main purpose is isolate the dredging material of lower quality inside the subaquatic cell which will be capped with better quality material. According VOGT (2009) the disposal can be in natural depressions in the seafloor, in borrow pits in the seafloor from mining operations (e.g., beach nourishment), or in specifically designed and constructed cells to contain the contaminated dredged material, which is exactly the case of the Piaçaguera Channel.

During the years 2016 and 2017, a CAD cell was built using different dredging methodologies and following strict technical and environmental control and monitoring standards, with the opening of the CAD being completed in June 2017.

Once the CAD was opened, it began to be filled with the dredged material of the Piaçaguera Channel. The first step of filling was finalized on December 2017, since then the deposited material is consolidating by its own weight for a few months until optimum density is reached so the second step of filling can be carried out and, subsequently, the CAD will be capped with suitable material to isolate the lower quality material disposed inside the CAD.

2. CRITERIA AND PREMISES ADOPTED

The CAD was designed to contain all material not suitable for disposal in the open sea in the Piaçaguera Channel, which in the first step of filling was equivalent to a volume of around 2,300,000.00 cubic meters.

Considering recommendations from USACE (2004), was adopted a bulking factor close 50.0%, and it was necessary to excavate a CAD with a useful volume of approximately 3,000,000.00 cubic meters. Figure 2 shows a typical section of the CAD.

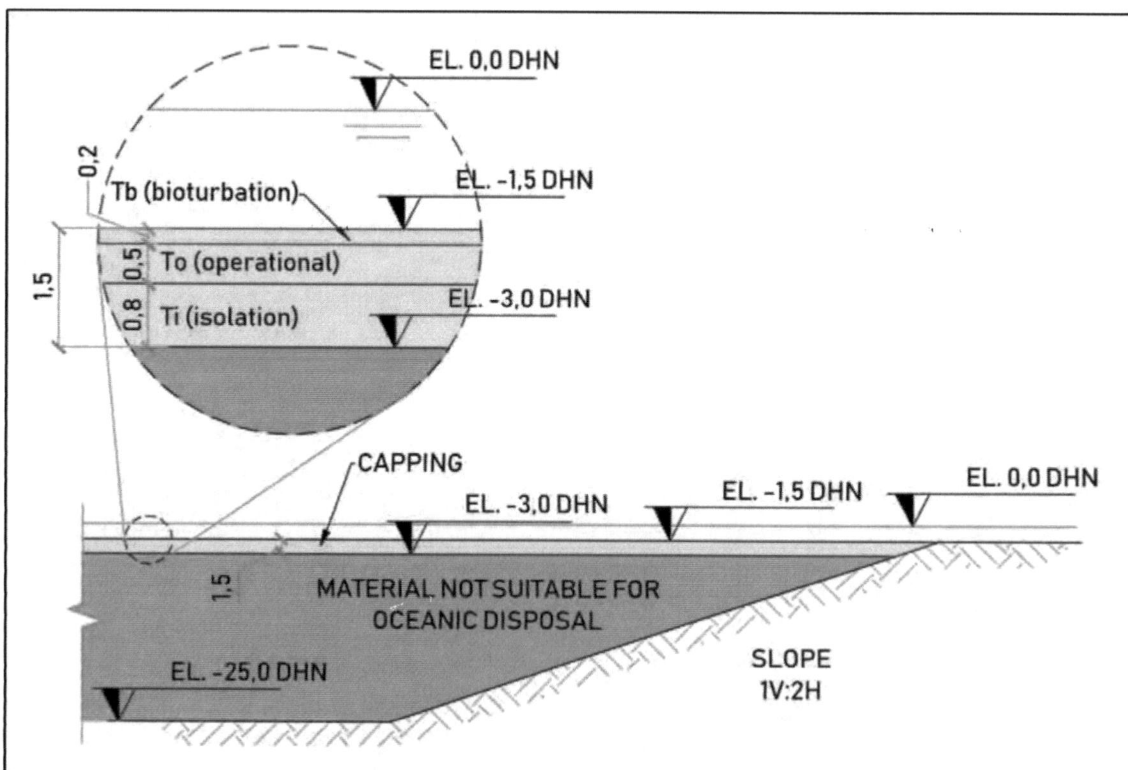

Figure 2: Typical section of the CAD.

Since the location of the CAD was a very shallow area (Figure 3) and to comply with the guidelines of USACE (1998) regarding the thickness of the cap of at least 1.5 meters, it was necessary to excavate another 500,000,00 cubic meters, totaling an excavation of 3,500,000.00 cubic meters for the CAD opening.

To open the CAD, because the depths of the site were very shallow, less than 2.0 meters deep (Figure 3), the excavations were initially carried out using a set of mechanical equipment formed by a clamshell operating with a bucket of 20.0 cubic meters and two barges of 2,800.00 cubic meters each. Monthly, about 200,000.00 cubic meters of material were dredged and disposal in the open sea in the ODP - Oceanic

Disposal Polygon where the good quality dredged material of the Port of Santos is usually disposed. This first phase ended approximately 4.0 months when 800,000.00 cubic meters were dredged.

The set of mechanical equipment was then replaced by two Trailing Suction Hopper Dredger – TSHD with approximately 10,000.00 cubic meters of hopper volume, which alternately excavated the CAD to a maximum depth of 25.0 meters (mean of 22.0 meters) and removed during six months approximately 2,700,000.00 cubic meters. Altogether, approximately 3,500,000.00 cubic meters were dredged in the opening phase of the CAD.

Figure 4 shows images of the equipment used during the CAD opening.

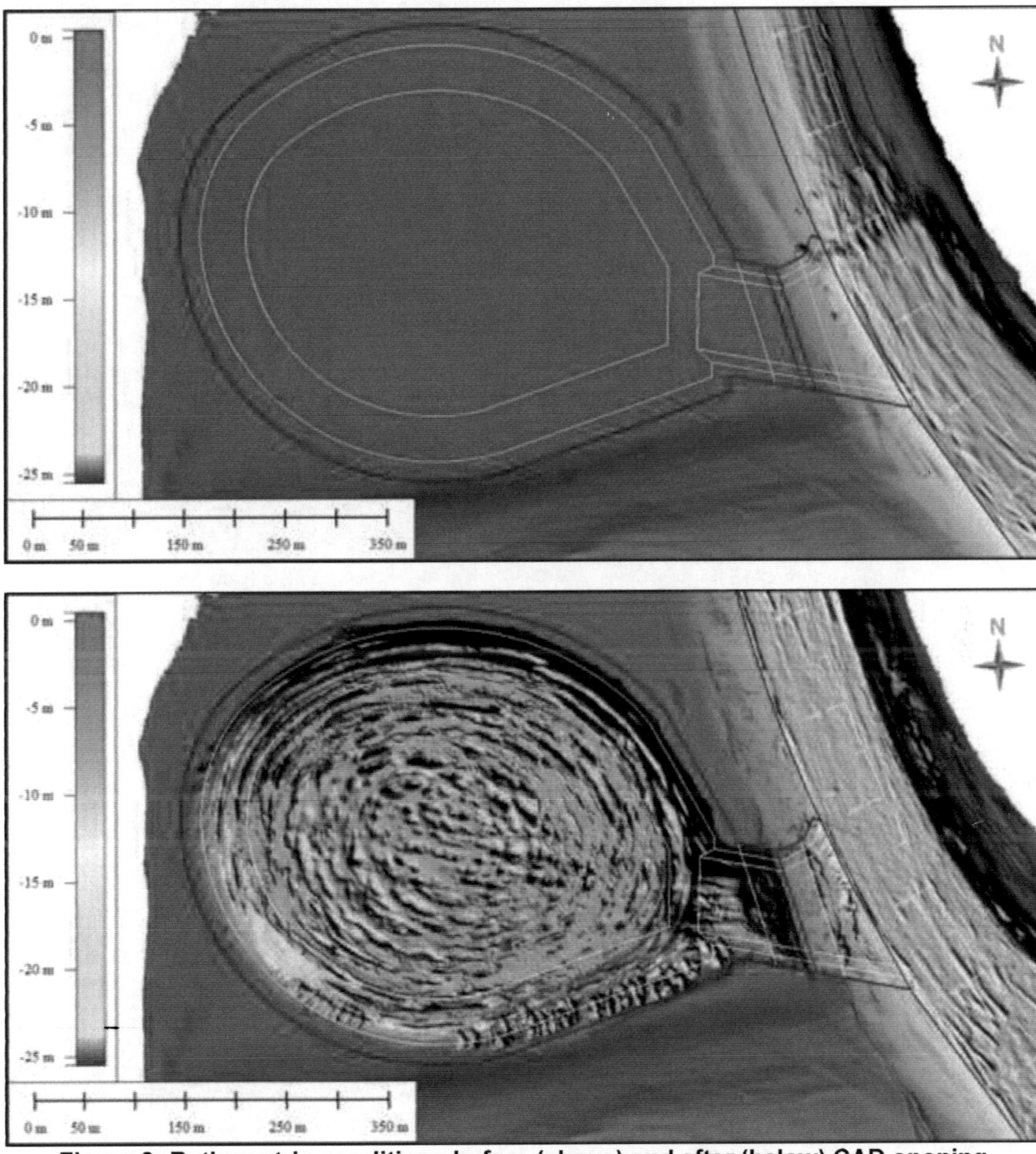

Figure 3: Bathymetric conditions before (above) and after (below) CAD opening.

Clamshell | Barge | TSHD Geopotes 15 | TSHD Ham 316

Figure 4: Equipment used during the CAD opening.

3. CONSTRUCTIVE METHODOLOGY

The implementation of the CAD was planned to take place in three distinct phases: i) opening, ii) filling (2 steps) and iii) capping. The first phase was completed.

The total opening of the CAD took place in June 2017 and soon after that began the first step of filling. To do so, before the beginning of the disposals, a silt curtain was implanted around the entire perimeter of the CAD to control the dispersion of suspended solids to the external ambient (Figure 5).

Silt curtains are flexible floating barriers that partially block the streams that carry the suspended solids, reducing the dispersion of these solids outside the perimeter protected by them. Such curtains are widely used in dredging works and the effectiveness of their use has already been evaluated and proven by different and independent laboratory and field studies, especially when the barrier is not open and when there is an "adaptive handling" in the positioning and dimensions (RADERMACHER et al, 2013). This was the case of the Piaçaguera Channel, where a specialized technical team was set up since the installation of the curtain, dedicated 24 hours/day to the handling and maintenance of the barrier.

During the adaptive handling of the silt curtains always when some trench is opened, the supervision team promptly held the barrier closing. To facilitate the anchoring and mooring of the barrier, some steel piles were also implanted around the entire cavity as shown in Figure 5.

Silt curtain around the CAD | Maintenance of the barrier

Figure 5: Silt curtain implanted around the entire perimeter of the CAD to control the dispersion of suspended solids.

For the dredging of the Piaçaguera Channel and CAD filling, one of the two TSHD used during the opening step was used. The dredge was excavated and pumped through a PEAD tube piping into the CAD (Figures 6 and 7).

Figure 6: Illustrative image of the material disposal method within CAD and detail (right) of connection between TSHD and floating piping.

Figure 7: TSHD performing pumping through PEAD piping to fill the CAD.

At the end of the tube was installed a diffuser with the objective of directing the fluidized material to the bottom by reducing the transport velocity. Its position was controlled through anchors. The objective of the adoption of the diffuser was to minimize the effects of turbulence during the descent and scattering of the sediment after the bottom plume collapse. This equipment has a controlled flow velocity and a conical shape, increasing the cross-sectional area near the end, which significantly decreases the velocity of the fluid, helping to dissipate the turbulence and minimize the dispersion of sediment during disposal. In addition, the diffuser had a geotextile skirt made in steel to guide the boom to a depth of approximately 10.0 meters, preventing the sediment from spreading on the CAD surface (Figure 8). This skirt had its depth adjusted as the level was raised during the filling of the CAD.

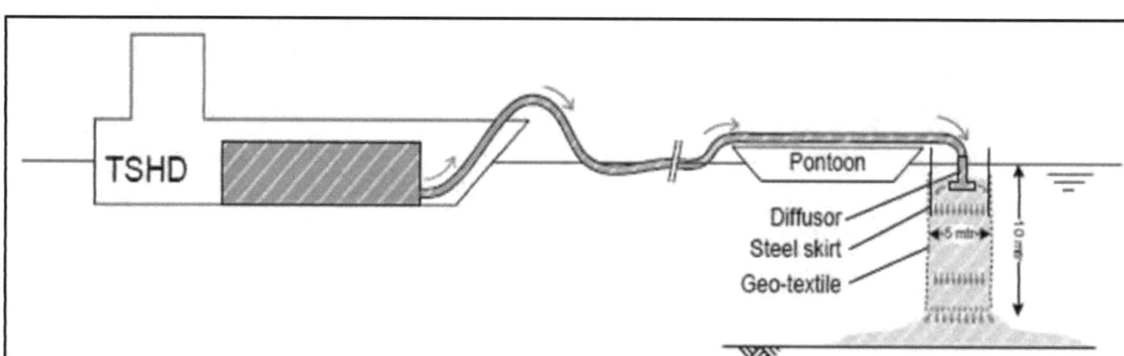

Figure 8: Illustrative image of the TSHD operating in combination with a diffuser.

The first step filling of the CAD and dredging of the Piaçaguera Channel occurred in December 2017. This step lasted 5.5 months and dredged about 2,300,000.00 cubic meters, which were fully dampened inside the CAD.

This first step of the CAD filling and channel dredging has been done to remove as much material as possible and to establish a depth of 13.5 meters along the channel to improve the conditions of navigability.

The second step of completion is expected to occur in 2018, when dredging of the channel will continue until depth of 14.5 meters is reached so the total channel cleanup occurs and, in the future, during maintenance dredging, will only be dredging of good quality sediment originating from natural channel silting that can then be dampened in the open sea.

Currently the material inside the CAD are in the process of consolidation by its own weight. Often bathymetric surveys and other tests are carried out with the objective of monitoring the consolidation process for the decision making on the best moment to conclude the dredging and perform the capping to definitive isolation of the material. In parallel, environmental controls are being carried out, which will be presented below.

4. MANAGEMENT AND CONTROLS

The management of the dredging works was performed to ensure that the opening and filling of the CAD as well as the dredging of the channel were executed as foreseen in the engineering project, obeying the contractual deadlines and prices and in accordance with the environmental permitting and governmental authorizations.

During the entire period of the work, a technical team was made of engineers, supervisors, HSE technicians, environmental experts and managers who checked and registered all the activities carried out both on land and on board, ensuring that dredging and dumping of material were carried out at the sites previously determined and authorized and that the activities were carried out to the highest HSE standards.

Among the technical controls established during the work, it is worth mentioning the online continuous registration by satellite navigation and positioning of the equipment allocated, time record and place of the beginning and end of the dredging and disposals, the frequent monitoring of the bathymetric conditions in the dredging areas (CAD and channel) and the dumping sites (ODP and CAD) and the slopes monitoring of the CAD area.

Considering that the CAD implementation has not yet been finalized being necessary to carry out the second step to fill and finally the capping, actually the monitoring of the material consolidation within the CAD is being carried out through bathymetric surveys, sample collections and analyzes, as well as *in situ* measurements of density.

As previously mentioned, a silt curtain was also implemented, which contributed to reduce the dispersion of suspended solids during disposals to fill the CAD. This barrier will continue to be installed until the end of the material capping.

Specifically, in relation to environmental controls, an extensive monitoring program was implemented, which involves the sampling and analysis of soils and water in the Piaçaguera Channel CAD and adjacent region, as well as monitoring fish, local fishing communities, birdlife monitoring, monitoring of vessel traffic and some other specific environmental programs and action.

5. LEARNED LESSONS AND FINAL CONSIDERATIONS

Comparing the planning with what was executed concludes that there was a good correlation. The dimensions designed for CAD, as well as the bulking factor and geotechnical slopes adopted were adequate for the type of material and the total volume of sediments to be dredged and confined.

The constructive methodology adopted, which employed two large hopper dredgers as a function of the need to perform the CAD opening phase rapidly and assertively, proved to be successful. Although the hopper dredges operated in a shallow and confined area, there were no significant interruptions even in the face of the difficulties of maneuverability and the need for precise cuts to obtain the depths and slopes designed for CAD. It should also be noted that the implantation of silt curtains and their adaptive management, as well as the diffuser at the end of the pipeline, shown adequate measures to control the dispersion of the turbidity plumes generated during the disposals.

The case of the Piaçaguera Channel broke paradigms where substantial volumes of sediments not suitable for the oceanic disposal were managed in a safe way allowing the implementation of more favorable and competitive navigation conditions along the Piaçaguera Channel and, above all, the establishment of better environmental quality standards, almost reducing the presence of contaminated sediments along the channel.

After completion of the filling and capping of the CAD and the consequent total cleanup of the channel, associated with several other social-environmental actions, it is hoped that the Piaçaguera Channel will present environmental quality much better than the current one.

6. REFERENCES

IADC – International Association of Dredging Contractors. (2010). Facts about Confined Disposal Facilities: An information update from the IADC. Number 1.

RADERMACHER, M, VAN DER GOOT, F, RIJKS, D.C, & DE WIT, L. (2013) The art of screening: effectiveness of silt screens. Terra et Aqua, Number 132, pp. 3-12.

USACE - US Army Corps of Engineers. (2004). Dredged material sedimentation study for Piaçaguera canal and Cubatao harbor, Brazil.

USACE - US Army Corps of Engineers. (1998). Guidance for subaqueous dredged material capping. Technical Report DOER-1. Vicksburg, MS.

RESILIENCE & ANTI-FRAGILITY OF THE NEW JERSEY STATE-MAINTAINED MARINE TRANSPORTATION SYSTEM

Matthew Lunemann[1], Jake McTavish[2], Genevieve Clifton[3], W. Scott Douglas[4], Witt Barlow[5], Jennifer Grenier[6], Michael Marano[7], and Martin Snow[8]

ABSTRACT

Following Superstorm Sandy, responsibility for maintenance and recovery of New Jersey's state-maintained navigation channels shifted to the New Jersey Department of Transportation's Office of Maritime Resources (NJDOT OMR). Immediately following the storm, NJDOT OMR implemented a system-wide data collection and prioritization program to restore the state-maintained marine transportation system's (MTS) channels and dredged material storage facilities to a state of good repair. In addition to the Port of New York & New Jersey, a multi-billion-dollar fishing industry, and hundreds of marinas, the NJ MTS consists of over 370 nautical kilometers of state-maintained channels and nearly 70 dredged material confined disposal facilities (CDFs). Bathymetric surveys and over 5,000 sediment cores were collected from 209 of NJ's 216 state-maintained channels within 5 months. To assist coordination of these surveys and sampling, OMR solicited the participation of numerous stakeholders, including the US Army Corps of Engineers, US Coast Guard, NJ State Police, and design consultants from WSP and Gahagan & Bryant Associates (GBA). To establish a sequential recovery plan for the NJ MTS, the state-maintained channels were prioritized based on economics, social importance, and other factors. After 5 years, the MTS recovery is still on-going, though significant progress has been made. The extent of impacts to navigation within the channels and dredged material storage capacity have been computed for pre-storm, post-storm, and current conditions using post-storm damage assessments, sediment core sample analysis, and bathymetric surveys. Analysis of these values for channels both singularly and system-wide, and for individual CDFs and system-wide dredged material storage capacity, portray not only the resilience but also the anti-fragility of the NJ MTS, as values have improved to better than pre-storm levels. The recovery efforts have also demonstrated a feasible link between resilience and sustainability by focusing on beneficial use of dredged material to restore damaged CDFs, restoration of habitat degraded from underwater borrow pits and restoration of degraded coastal marsh via placement of dredged material. A case study of a multi-agency cooperative marsh restoration pilot program in Fortescue, New Jersey is provided, which demonstrates significant additional potential for enhancing the resilience of the NJ MTS. As a result of the post-storm prioritization and recovery planning, a shift in public sentiment regarding dredging, OMR's focus on establishing a sustainable dredging program, and availability of recovery funding, the NJ state-maintained MTS is on pace to not simply recover to pre-storm conditions, but to recover to a higher level of resilience than before the storm.

Keywords: Recovery planning, dredged material, beneficial use, marsh enhancement, thin-layer placement, confined disposal facility.

---

[1] Geotechnical Department Manager, WSP, 4 Penn Center, 1600 JFK Boulevard, Suite 510, Philadelphia, Pennsylvania 19103, USA, T: 609-512-3663, Email: Matthew.Lunemann@wsp.com.

[2] Civil Engineer, Gahagan & Bryant Associates (GBA), 9008-O Yellow Brick Road, Baltimore, Maryland 21237-5608, USA, T: 410-682-5595, Email: jamctavish@gba-inc.com

[3] Manager, New Jersey Department of Transportation Office of Maritime Resources (NJDOT OMR), 1035 Parkway Avenue, Trenton, New Jersey 08625, USA, T: 609-530-4770, Email: Genevieve.Clifton@dot.nj.gov

[4] Dredging Program Manager, NJDOT OMR, 1035 Parkway Avenue, Trenton, New Jersey, 08625, USA, T: 609-530-4770, Email: Scott.Douglas@dot.nj.gov.

[5] Senior Vice President, GBA, 9008-O Yellow Brick Road, Baltimore, Maryland 21237-5608, USA, T: 410-682-5595, Email: wbarlow@gba-inc.com

[6] Project Manager, WSP, 2000 Lenox Drive, Lawrenceville, New Jersey 08527, USA, T: 609-512-3575, Email: Jenn.Grenier@wsp.com

[7] Senior Supervising Civil Engineer, WSP, 2000 Lenox Drive, Lawrenceville, New Jersey, 08527, USA, T: 609-512-3516, Email: Michael.Marano@wsp.com.

[8] Vice President, GBA, 9008-O Yellow Brick Road, Baltimore, Maryland 21237-5608, USA, T: 410-682-5595, Email: mrsnow@gba-inc.com

INTRODUCTION

New Jersey is a maritime state, surrounded by water on three sides. A significant portion of New Jersey's economy is dependent on its maritime infrastructure. New Jersey's Marine Transportation System can be divided into three major regions: the NY/NJ Harbor, the Delaware River, and the Atlantic Shore. Within these three regions are over 930 km of engineered waterway, thousands of berths, docks and ramps, as well as two internationally significant port complexes with associated terminals and goods handling facilities. Overall, New Jersey's MTS supports an economic engine worth over US$50 billion annually ranging in activity from tourism and recreation to commercial fishing to international trade. Millions of cubic meters of sand, silt and gravel are transported into this channel network by Mother Nature every year and must be removed by maintenance dredging.

On the evening of October 29, 2012, a post tropical cyclone made landfall in Brigantine, NJ, just north of Atlantic City. This storm, named by the media "Superstorm Sandy", was the largest Atlantic storm on record with winds extending over 1800 kilometers. Superstorm Sandy brought unprecedented storm surge (4.23 meters), waves (9.9 meters) and winds in excess of 125 kilometers per hour to the mid-Atlantic region, causing significant damage to homes, businesses and infrastructure. At the peak of the storm, 2.6 million NJ customers were without power and twelve New Jerseyans lost their lives. Over 160 kilometers of shoreline were severely eroded, most of it in New Jersey and Long Island, New York. In New Jersey alone, damage exceeded US$36 billion, making Superstorm Sandy the costliest storm in New Jersey history.

Resilience is defined by the National Academy of Science as "the ability to prepare and plan for, absorb, recover from, and more successfully adapt to adverse events. PIANC Task Group 193 follows a similar definition of "the capacity to anticipate and plan for disruptions, resist loss in operations and/or absorb the impact of disturbances, rapidly recover afterwards, and adapt to short- and long-term stressors, changing conditions and constraints". Essentially, resilience is an iterative process with four distinct stages: pre-planning before a disturbance, resisting loss during a disturbance, recovery after a disturbance, and adapting for the next occurrence. In certain circumstances, a system may exhibit anti-fragility, or the ability to recover stronger and more resilient than pre-disturbance levels.

This paper highlights the NJ state-maintained MTS (NJ MTS) recovery and adaptation phases as well as recovery planning conducted during the initial stages of the recovery. In their recovery efforts, the New Jersey Department of Transportation's Office of Maritime Resources (OMR) has begun the iterative resilience process, resulting in a NJ MTS exhibiting anti-fragility and growing resilience.

RECOVERY PLANNING & PRIORITIZATION

Immediately following the storm, significant efforts were undertaken to evaluate the damage and begin recovery efforts. While damage to the ports on the Delaware River and in the NY/NJ Harbor was handled by their respective Port Authorities, damage to the Atlantic Shore infrastructure fell to the State's Departments of Environmental Protection (NJDEP) and Transportation (NJDOT). Historically, the NJDEP's Bureau of Coastal Engineering was responsible for maintenance dredging of the state navigation channels. However, the extent of the damage from the storm and the need for rapid deployment of multiple contractors led the administration to ask for the NJDOT's Office of Maritime Resources (OMR) to take charge of the channel recovery effort. NJDOT immediately initiated a cooperative interagency workgroup to perform the necessary condition surveys and associated evaluation needed to permit and procure dredging and debris removal contracts. Representatives of multiple State and Federal agencies were partners in the process including the NJDEP, NJ Office of Emergency Management, US Army Corps of Engineers, US Coast Guard, as well as engineers from Parsons Brinckerhoff (now WSP), Gahagan and Bryant Associates, and Dewberry Engineers.

Field reconnaissance and side scan sonar was used to locate, and eventually remove, large debris and wreckage in the channels, but the greatest impact was from sediment washed into the navigation channels. In order to define this, an unprecedented system-wide bathymetric survey was undertaken. Using single beam sonar technology, over 370 kilometers of shallow draft navigation channels (209 individual channels) were surveyed in 4 months.

Debris Removal

As side scan survey data was processed, a detailed review of each channel-specific dataset was performed. The number of identified targets in each channel was documented and sorted by channel, type, and NJDOT Region. Geo-referenced mosaics were generated, and identified debris targets were geocoded to their coordinate position and used to generate a statewide database of storm-related channel data. To assist the debris removal efforts, each target reference included a series of attributes, including width, height, length, shadow, and a classification/description based on a visual review of the target area. These data were documented as part of a GIS-ready shapefile as well as a channel-specific summary document that included overall channel scan imagery, a summary table of targets, and a target-specific summary table.

The resultant survey provided data for what ultimately became the largest waterway debris removal project in United States history. Nearly 5000 pieces of debris, including 36 vessels and 20 dock sections, were identified and located for NJDEP debris removal contractors. Over 450 of these pieces of debris were large enough to create a hazard to navigation and were either removed or marked for later removal.

Channel Prioritization

In addition to the debris, approximately 2.3 million cubic meters of sediment were identified within the state-maintained channels above the design depth. To derive this quantity, single beam bathymetric surveys were performed to confirm channel depths and ultimately identify shoal locations and material quantities. Survey lines consisted of cross-sectional transects run perpendicular to the channel centerlines at 30-meter intervals, as well as two longitudinal tracks run parallel to the channel centerlines at a 7.6-meter offset either side.

After acquisition, data was post-processed and presented on plan sheets to show preliminary alignment and depths. These bathymetric plans were reviewed by State engineers to determine the best centerline location to maximize navigability and minimize dredging requirements. This strategy is believed to have reduced the number and/or extent of shoals by as much as 10%.

The evaluation also served multiple other purposes: to review the physical condition of the channels, including shoals and debris targets; to prioritize remediation efforts through an evaluation of the economic value and usage of specific channels, and; to repair the channels in order to protect the public from hazardous conditions resulting from the storm.

Once the final alignments were determined, the data was imported to engineering and modeling software. Design template alignments, illustrated in Figure 1, were constructed in Terramodel, a product of Trimble, and average end area shoal quantities were calculated between the templates and surveyed existing ground surfaces. This graphic represents the dredge envelope (green) and allowable overdredge (blue). The dredge envelope represents the channel design depth for its entire width, plus 3H:1V side slopes extended upward from the outer extents until they intercept the existing bathymetry. Overdredge represents the allowable area that can be dredged beyond the proposed depth (for this project, 0.3 meters) to ensure that at a minimum, the proposed depth is met throughout the entire channel. In addition, surface-to-surface DTM quantities were calculated for comparison as quality control.

Shoaling calculations were performed and reported for each 30-meter section of channel, separated into left and right quadrants in an Excel spreadsheet. Shoaling in each channel was reported using both a graphical display for the entire channel that showed the location of each shoaled quadrant and with a summary report that totaled the volume in each quadrant by both minimum and maximum dredge depth (design depth plus 0.3 meters).

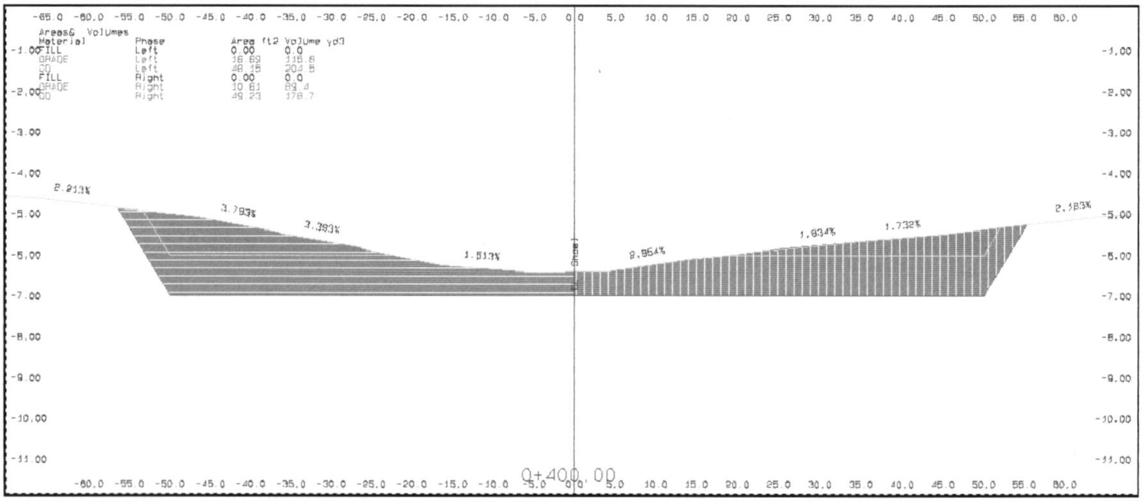

Figure 1 – Typical cross section of data and design template during average end area calculations showing grade and overdepth regions

To further define the status of each channel, the team also made a determination of navigability through each shoaled area. Each shoal was "ranked" as follows:

Minor: Reasonably navigable (at or near authorized depth.)

Moderate: Navigable, but with caution; may be impassable for some vessels at low tide.

Severe: Navigability is severely impaired for many vessels even at high tide; dangerous conditions may be present at some tide stages for some vessels.

An overall rank was given to each channel based on the location of the shoals within the channel relative to the location of users and the orientation of the channel to the entire channel network.

Once the debris was removed and each shoal that was considered a hazard to navigation marked, the massive task of removing the 2.3 million cubic meters of sediment was initiated. In order to prioritize this task, OMR initially evaluated projects based on their readiness to proceed and dredged material disposal availability in the following order of precedence:

1) Shovel-ready projects already permitted from the pre-Storm program
2) Projects with existing dredged material disposal sites with capacity available
3) Projects with dredged material disposal potential
4) No dredged material disposal options

To rank projects within the above groupings, OMR considered the economic benefit and severity of shoaling of specific channels to provide a form of cost-benefit and need assessments. A series of Economic Value and Channel Usage Ratings were developed to aid in the assessment of the impact of each channel on the local economy as well as the importance of each channel on statewide navigational/travel patterns. These need-based and cost-benefit assessments were especially useful where more time and effort was required to initiate a project, such as channels requiring development of dredged material disposal options and/or purchasing property for disposal sites.

Assessing Storm Impacts

Using the final approved channel centerline and limits, a team of maritime professionals (engineers, marine scientists and coastal planners) from the NJDOT Office of Maritime Resources and the NJDEP Bureau of Coastal Engineering reviewed available data to evaluate whether shoaling might be attributable to Sandy. Where possible, the determination of channel impacts was made based on comparison of pre-storm and post-storm bathymetry. However, as this channel bathymetry was only available for 49 channels at the time of the storm, OMR and their team reviewed approximately 5,000 core samples taken from within 116 channels after the storm. The cores were inspected for debris, composition, and other characteristics to assess storm-produced sediment versus pre-storm sediment. Where both pre- and post-storm bathymetry and core sampling data was not available, a determination based on best professional judgment was made, including the following considerations:

- Input from local boaters
- Location of the waterway and its orientation relative to the storm path
- Surrounding landforms
- Fetch (amount of open water near the channel)
- Proximity of the channel to ocean inlets
- Bathymetry of surrounding area (evidence of sand waves)
- Direct field observations of storm damage (e.g. landward sand deposition, wave damage)
- Navigation incident reports from boaters, marina owners, local governments, and the Marine Services Bureau of the New Jersey State Police
- Storm surge mapping
- Historical channel maintenance records

For simplicity, channels with similar impacts were grouped by the geographic regional clusters. This regional assessment, along with available channel specific information allowed the team to label each channel as "yes", "no" or "maybe" with respect to Sandy impacts. This information was used to prioritize storm recovery work, negotiate for FEMA reimbursement, and to prioritize future dredging efforts statewide.

Multi-Party Coordination

Critical to the recovery was clearing or marking all channels before boating season was in full swing (May 2013). A response of this magnitude required a multitude of agencies to be engaged on a variety of levels and times. To achieve this massive task in an aggressively set timeframe, NJDOT and NJDEP created a multidisciplinary team of consultants and state and local agencies, each with defined roles in the recovery process. The consultant team developed a schedule for collection, processing and reporting of sidescan sonar data and bathymetric survey data. The team also utilized a series of management tools, including daily coordination conference calls and ProjectSolve2: an Internet-based file sharing tool where project staff could access and review the numerous files generated for each channel.

In order to clear and/or mark hazards to navigation quickly and manage the massive data collection and reporting, the team developed a series of processes and tools for effective and efficient communication. Each channel was given a unique identifier and a database for reporting was created. This database included channel name and number, date of sidescan survey, date of sidescan processing, number of targets identified, date the sidescan data was uploaded to the ProjectSolve2 site, date of hydrographic survey, date of hydrographic processing, date of shoaling reports uploaded to ProjectSolve2, and number of shoals identified.

Each day, data was collected, processed, and reported for marking and/or clearing debris. The team held daily calls with the NJDEP (debris removal team), NJDOT OMR, NJ State Police, NJ Office of Emergency Management (OEM), FEMA, and US Coast Guard. On these calls, the team reported on:

- Which channels were completed that day and where they would be the next day
- Progress on debris identification,
- Progress on debris marking or removal

Data was uploaded to ProjectSolve2 and available for NJDEP to utilize for debris removal, ATON and OEM for marking hazards to navigation and US Coast Guard and State Police for notice to mariners. The constant communication amongst project team members and timely response of all project team members allowed the project to be completed on time in only 4 months.

RESILIENCE & ANTI-FRAGILITY OF THE STATE-MAINTAINED MARINE TRANSPORTATION SYSTEM

As a result of the recovery planning, multi-agency coordination, and integration of team members, the OMR team quickly utilized the collected data to begin the process of restoring the state-maintained navigable channels and to restore or identify dredged material storage options. The effective planning prior to the start of recovery efforts has resulted in anti-fragility and increased resilience of the NJ MTS for the navigable channels and dredged material storage capacity, both of which contribute directly to the rate of system recovery.

Navigable Channels

From analysis of collected survey data, over 2.3 million cubic meters of sediment were identified above the design depth across the state-maintained MTS. Over half of the 209 channels were significantly shoaled, and 34 percent were severely shoaled or impassable. Comparison to recent bathymetry taken prior to the storm showed that the storm had deposited a significant amount of material in some places. Unfortunately, pre-storm bathymetry was available for only 49 channels. In order to determine the storm's impact on the rest of the system, over 5000 cores were taken in shoaled areas down to authorized depth. Geologists examined the cores for signs of recent deposition in the form of vegetative debris, marine organisms and consolidation (or lack thereof). In the final analysis, as presented by Marano, M. et al. (2017), about 27 percent of the sediment could be attributed to Superstorm Sandy, or 610,000 cubic meters.

It is important to note that this analysis is limited to the State channel network, and does not include the needs of the myriad of local access channels, marinas, lagoon communities and private berths. Although there is no published study that defines the private and local need for dredging, our experience tells us that the numbers are even higher, mostly due to long neglect of maintenance. Nor are we including the more than 200 kilometers of Federal channels and inlets that serve the Atlantic coast of New Jersey, or the Port of New York & New Jersey for which resilience was previously reported by Wakeman, T.H. & Miller, J. (2013) and Wakeman, T.H. et al. (2015).

Based on review of bathymetric surveys performed in 2013 and 2015, natural sedimentation results in approximately 2 channels dropping out of a state of good repair each year and in two channels going from moderate to severe impacts to navigability. This natural diminishing of MTS operability is indicated by the baseline in red in Figure 2. When comparing to the post-Sandy baseline of MTS impacts to navigation in Figure 2, the number of channels in a state of good repair (i.e. minor to no impact) has exceeded pre-storm levels as of January 2018, demonstrating anti-fragility of the system.

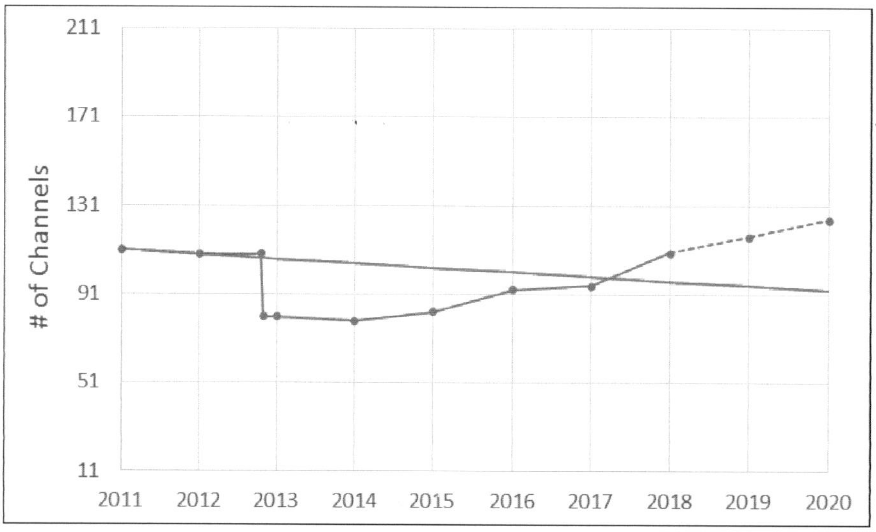

Figure 2 – Number of Channels in Good Repair with Baseline Post-Storm Trend (in red)

As noted in Figure 2, planned projects or those currently on-going will continue to improve the state-maintained MTS and the anti-fragile response hints at a higher rate of resilience expected following the next storm. Due to the recovery program and post-storm planning, 35 shoaled channels have been restored to a state of good repair as of January 2018. In all, 52 percent of channels are now in a state of good repair (up from 38% post-storm), with only 20 percent of channels experiencing severe impacts to navigability (down from 34% post-storm). OMR's recovery efforts to date have provided improved navigability to approximately 58 marinas, 69 businesses, 1 ferry terminal, nearly 200 commercial vessels, and over 2,000 residential docks.

While the loss in functionality shown in Figure 2 indicates a low to moderate risk and low resilience based on characteristic relationships presented by Linkov, I. et al. (2014), the measured resilience is reflective of the pre-storm funding levels and lack of available storage capacity for fine-grained sediments in the storm's aftermath. To combat this low resilience and improve recovery speeds following the next storm, OMR has focused on several dredged material management strategies as part of the on-going recovery efforts.

Dredged Material Storage

Equally critical to the restoration of storm-impacted navigable channels, dredged material storage of the NJ MTS was significantly impacted as a result of past-use and storm damage. In an ideally resilient system, sufficient storage capacity would be available to accommodate projected dredging needs following a storm in addition to maintenance dredging needs. In such a system, channel restoration could begin immediately following the storm and result in a quicker recovery and increased system resilience. As such, a direct correlation exists between the amount of available dredged material storage capacity and the resilience of marine transportation systems.

Confined Disposal Facilities

Historically, coastal New Jersey navigational dredging is performed with small (0.25 m to 0.45 m) cutterhead pipeline dredges. While dredged material processing and other dewatering methods, e.g. filter press, are available in New Jersey, dredged material has predominantly been managed in upland disposal sites or on beaches, depending on grain size. Material that is greater than 90 percent sand is suitable for bathing beaches, while material greater than 75 percent sand is potentially useable on non-bathing beaches. On a systemwide basis, about 20 percent of the maintenance material is coarse-grained (75% sand or greater), and 80 percent is fine-grained.

A recent survey of the state by NJDOT revealed over 45 historical dredged material placement sites along the east coast of NJ (not including beaches) ranging from free pump unconfined disposal to fully engineered confined disposal facilities (CDFs) with engineered level control structures. Many of these CDFs are small, with potential air capacities of less than 75,000 cubic meters. This is probably not an exhaustive list since many upland sites have been developed or have become unrecognizable over time. Regardless, not all of the sites are permitted for use due either to environmental regulations or having passed into private ownership. At the time of the storm, there were 25 useable CDFs along the state's east coast, with approximately 1.4 million cubic meters of air capacity. This would not be close to enough to manage the 2.3 million cubic meters of sediment within the MTS let alone the 610,000 cubic meters of sediment from the storm when accounting for operating capacity and applying swell factors to these in-situ sediment volumes. To compound matters, at least some of the capacity would need to be reserved for the Federal Intracoastal Waterway (ICWW), due to a long-standing agreement between the US Army Corps of Engineers (USACE) and the State.

Figure 3 shows the locations of these CDFs broken out by region.

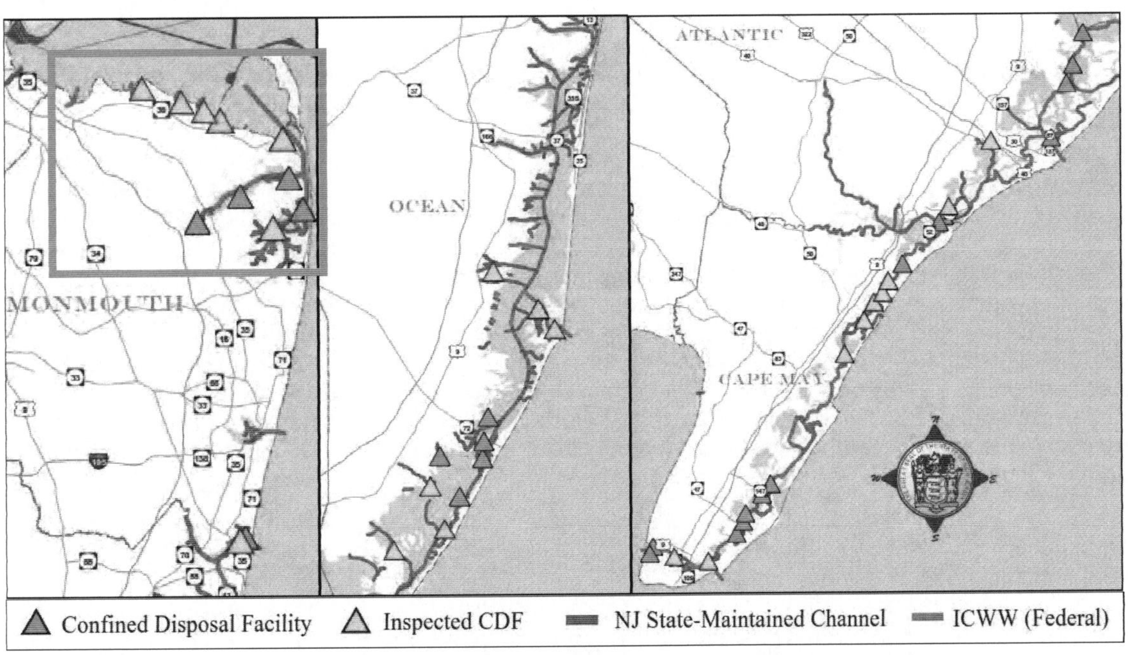

Figure 3. Locations of Inspected CDFs along Raritan Bay & ICWW: Raritan Bayshore (boxed) and Monmouth Rivers (Left), Barnegat Bay (Center), and Atlantic & Cape May Counties (Right).

To begin the rebuilding of the state channel network, considerable efforts were put into restoring capacity in CDFs. This included evaluating remaining capacity as well as identifying those facilities damaged by the storm and making appropriate repairs. These efforts included on-site damage assessments of 22 CDFs in the system, revealing significant damage at 11 of these CDFs which could not be used without likely breaching the confining dikes.

Based on computed estimates of available CDF capacities systemwide, the CDFs in the NJ MTS had approximately 1.4 million cubic meters of air capacity available before the storm, dropping to 1.2 million cubic meters of air capacity following the storm due to CDF damage.

Through a continuous program of emergency and maintenance dredging and CDF rehabilitation, a significant restoration in available CDF air capacity has been achieved with planned projects in 2018 projected to exceed pre-storm capacity levels systemwide, as shown in Figure 4 (left). In general, CDFs have been restored only when needed for dredging, as the priority has been on restoring navigability in the channels, resulting in the relatively flat trend indicated in Figure 4 (left) until planned CDF repairs are carried out in 2018. However, as indicated in Figure 4 (right), CDF dikes have continued to be raised systemwide resulting in an increased potential air capacity after removal of stored dredged material. While not all damaged CDFs have been repaired, potential CDF air capacity is expected to reach and exceed pre-storm levels in 2018. Taking the system as a whole, potential CDF air capacity has exhibited a low to moderate risk and low to moderate resilience based on the conceptual resilience relationships published by Linkov, I. et al. (2014), as indicated by Figure 4 (right). The storm-induced loss of functionality was sharp but a low percentage of the overall system capacity, with a variable recovery rate.

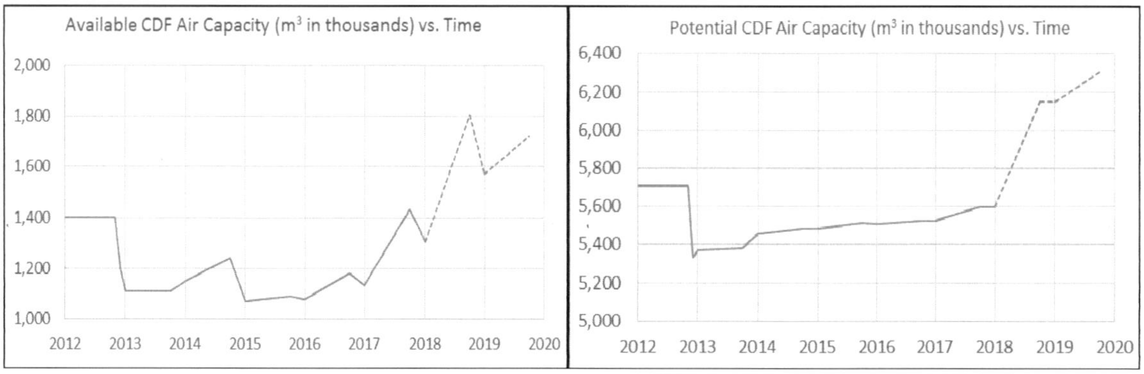

Figure 4 – Available CDF Air Capacity (left) and Potential CDF Air Capacity (right) over Time for the New Jersey State-Maintained MTS

Looking at all CDFs shown in Figure 3, a total air capacity of approximately 6.5 million cubic meters could theoretically be achieved if all accessible CDFs were repaired and rehabilitated using all on-site dredged material, without consideration for elevating the confining dikes. Roughly 1 million cubic meters of additional capacity could be achieved by elevating all confining dikes by 0.6 meters. However, while useful, the systemwide capacity can be deceiving, as the sharp but low percentage drop in overall potential air capacity averages out impacts across both low impacted and high impacted areas and is not indicative of severe regional impacts. In the case of the NJ MTS, the potential CDF air capacity in Figure 4 (right) was skewed higher due to the large percentage of total available capacity in Cape May County, which was less severely impacted by the storm than the Raritan Bayshore region. As a result, looking at storage capacity on a regional level returns a more representative assessment of the system resilience, as dredging in NJ is typically restricted to regional dredged material storage options.

This capacity loss from Superstorm Sandy was especially apparent in the Raritan Bayshore region of the state, which was hit the hardest from the storm and experienced a storm surge more than 2.8 meters above mean high water elevation.

Raritan Bayshore Region

Along the Raritan Bayshore (boxed in Figure 3), 4 CDFs were in locations that could be utilized for recovery efforts immediately following the storm; however, one was severely damaged by the storm, two were nearly full, and one had no available capacity. As such, only 14,500 cubic meters of air capacity was available immediately after the storm. The Raritan Bayshore region received a high priority from OMR following the storm, and air capacity was improved via elevating and emptying of 4 existing CDFs and construction of a new CDF at Leonardo. All increases in CDF dike heights were accomplished using dredged material on-site or from nearby CDFs. In the case of the N61 CDF, additional material to raise and repair the dikes was obtained via excavation and transport of roughly 38,000 cubic meters of dredged material from the Atlantic Highlands CDF and the Monmouth Cove Marina CDF, increasing capacity of 3 CDFs at one time. These projects not only supported dredging of 8 severely impacted channels, but also resulted in improvement to CDF air capacity that well exceeded the pre-storm levels, as indicated in Figure 5. Accounting for current available CDF air capacity (Figure 5 left), the post-use available air capacity has leveled off at around 100,000 cubic meters. Assuming future storage potential after removal of dredged material, the potential air capacity in the region has increased by roughly 80 percent from pre-storm levels as indicated in Figure 5 (right). Both measures of dredged material storage capacity show anti-fragility of the system directly resulting from the beneficial use of on-site dredged material to increase CDF dike heights. Unlike the navigable channels, the potential dredged material storage capacity in the Raritan Bayshore region demonstrated high risk and high resilience based on the relationships published by Linkov, I. et al. (2014) with a sharp, significant loss of functionality followed by rapid recovery. This higher resilience and anti-fragility is a direct result of OMR's prioritization of this region during recovery planning.

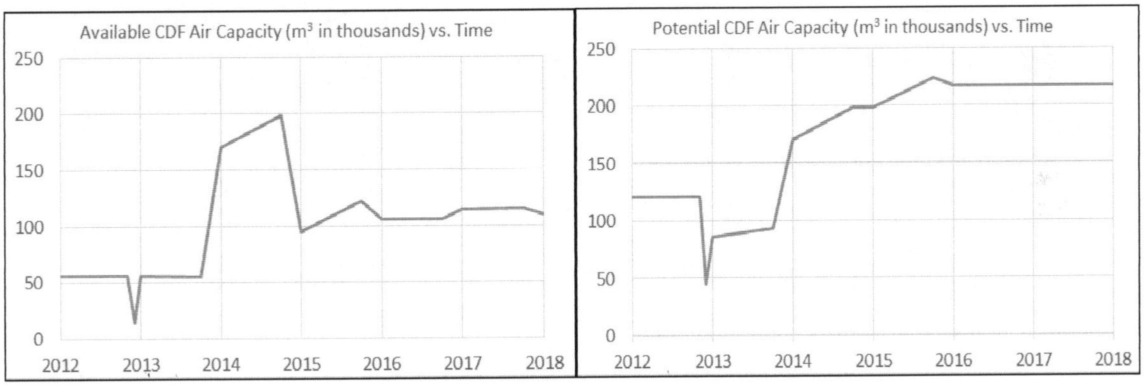

Figure 5 – Available CDF Air Capacity (left) and Potential CDF Air Capacity (right) over Time for the Raritan Bayshore Region

This increased capacity also allowed a Federal channel providing access for an interstate Ferry service to be dredged. To date, approximately 145,000 cubic meters of sediment has been dredged and placed into these CDFs from 7 channels. Currently, negotiations are underway to restore a historical CDF on private property in the Raritan Bayshore region. This CDF will provide support for a second Federal channel and 23 State channels (23.6 km total length), including one 3.4-km-long channel providing access for 5 marinas.

With additional CDF rehabilitation projects planned for the next several years, further improvement in dredged material storage capacity is expected. In many cases, hazard mitigation in the form of revetments or dike placement above the anticipated storm surge elevation will work to lessen damage and improve recovery time following the next major storm. As a result, increased resilience can be expected of the NJ MTS following the next major storm, and the data presented herein may be used as a benchmark upon which to measure the effectiveness of resilience improvements and anti-fragility of the NJ MTS.

Going forward, plans are underway to construct more CDFs along the beachfront in Keansburg, NJ where the CDFs will be integrated into the existing dune system to improve coastal resilience and combat sea level rise, providing a win-win by boosting resilience of both coastal communities and the NJ MTS.

Case Study: Dredging of Waackaack Creek & Thorns Creek and Keansburg CDF

The biggest challenge with dredging in New Jersey is the availability of placement sites. The placement of sand on the beach and/or restoring or creating protective dunes is an excellent example of the dual benefits of dredging and coastline resiliency provided by this project. The channels were cleared to their permitted depth and the dredged material was used to protect the coastline from tidal surges and flooding.

Another concept that serves this dual purpose is the construction of CDFs along a linear alignment parallel to the coastline (see Figure 6). This concept allows for the placement of more fine-grained material adjacent to the existing beaches. After placement, a filled CDF will provide more stability to the shoreline than the existing beaches which are more susceptible to erosion during storm events. A dune made on top of a CDF filled with fine-grained material would be more stable than a dune made solely of sand, since fine-grained material has more natural cohesion. The CDF constructed in the borough of Keansburg in Monmouth County, NJ is precisely this application (see Section A-A in Figure 7).

Figure 6 - Construction of beach-fill berm

The borough of Keansburg in the Raritan Bayshore region of New Jersey was severely impacted by Superstorm Sandy. In Keansburg, the US Army Corps of Engineers created a flood protection levee and installed a floodgate on Waackaack (pronounced Way cake) Creek in the late 1960's. The floodgate separated the inland portion of Waackaack Creek and Thorns Creek from the outer segment of Waackaack Creek that extends from the floodgate to the NJ Intracoastal Waterway (see Figure 7). The nature of the material to be dredged consisted of predominantly beach-quality sand in the outer channel and fine-grained material inside the floodgate.

Figure 7 – Project Map with Section A-A of the Keansburg CDF

An upland CDF already existed less than 60 meters from the mean high water line, but it was already filled to capacity with dredged material from the previous maintenance cycle. NJDOT approached the borough with the prospect of removing the dredged material for the construction of another coastal CDF in a nearby Monmouth County town. The borough did not accept this proposal and viewed this coastal CDF and the material in it as an added protection to their town from the next storm event.

The design team accepted this constraint as a positive and built upon it. The CDF was designed to remain within the original permitted footprint, while using the existing interior material to raise the confining berms, thereby creating capacity. This allowed for the placement of the fine-grained dredged material into the raised portion of the CDF and provided the town with an even larger natural barrier to Raritan Bay storm surges. The coarse-grained material from outside the floodgate was designed into a dune-like feature simply by extending the top of the confining berm elevation waterward and transitioning to the lower existing beach grade at a 3:1 horizontal to vertical slope. The slope met the existing beach grade well upland of the mean high water elevation line.

The dual benefits of the long-awaited dredging of Waackaack and Thorns Creeks, combined with a naturally hardened section of coastline, was deemed such a success that the borough requested the State construct additional, similar CDFs along the coastline adjacent to the existing Keansburg CDF.

Based on possible footprints for two additional CDFs built into the dune system, an additional 85,000 to 115,000 cubic meters of storage capacity could be gained for the Raritan Bayshore region. With such an increase in capacity, the MTS recovery in the region following a future storm would be greatly accelerated, and the CDF dune system would provide additional protection to the near-shore properties behind the CDFs. OMR's collaboration with local stakeholders and planning for future storms has directly resulted in increased MTS resilience and provides a win-win for both OMR and Keansburg. Coordination with state regulatory agencies has opened other opportunities for dredged material management, including confined aquatic disposal to restore or enhance underwater habitat.

Confined Aquatic Disposal – Dredged Holes

Confined aquatic disposal (CAD) facilities are defined as below water depressions in the existing ground that serve as open water dredged material placement areas. A CAD pit, or cell, may be formed from either natural events or intentional subaqueous construction. Suitable below water depressions, termed "dredged holes" or "borrow pits" exist throughout the bay and estuary system in New Jersey and have the potential to be used as a dredged material management solution to combat shoaling within the New Jersey state-maintained channels.

CAD facilities satisfy the same need as upland confined disposal facilities (CDF) in terms of storing dredged material. However, the two types of disposal facilities differ in several categories such as habitat, site accessibility, and material placement methodology. Like upland sites, CAD facilities are able to accept unsuitable, or contaminated material by use of a controlled cover, or cap. The cap consists of coarse-grained material and serves to ensure minimal loss of material from within the hole after the placement has occurred. A major advantage to the use of an existing cell as a CAD facility is that minimal to no construction is required upfront. Further advantages over other conventional placement methods are that these sites allow for either hydraulic pipeline discharge or mechanical placement, require minimal equipment needs, and are often located near dredging sites. A specific disadvantage to a CAD facility stems from the need of the cap when placing fine and contaminated sediments. In order to reduce the risk of resuspension and loss of dredged material from within the hole, the cap may be required immediately following a placement event, thus dictating the potential need for an interim cap in the event of a disruption of placement, such as a seasonal dredging permit restriction. Further disadvantages of these sites include the need to prevent loss of sediments during transport and placement, and potential difficulty obtaining agency approval.

A number of these CAD facilities were inadvertently created in New Jersey waterways when sand was excavated for upland construction in the early to mid-1900s. A January 2015 study titled 'State Channel Maintenance Capacity' performed by NJDOT-OMR, Richard Stockton College of New Jersey Coastal Research Center (CRC), and Ocean and Coastal Consultants (OCC) investigated the presence of potential facilities and identified a total of 122 dredged hole features along the NJ MTS. Amongst these 122 features, 10 were identified as candidate sites after field personnel measured depths greater than 5.5 meters below the surface. Upon establishing these candidate sites, hydrographic surveys were performed and yielded volumes for dredged material containment of over 1,720,000 cubic meters to elevations at or below -3.7 meters mean low water (MLW).

Amongst these identified candidate sites, is a feature shown in Figure 8, located in Brick Township, Ocean County denoted as Dredged Hole #18 (DH18). Water quality samples retrieved by CRC in 2013-2014 indicated that the benthic habitat within this feature is azoic due to extended periods of hypoxic conditions (dissolved oxygen levels of less than 2 mg/L). In its present state, DH18 is considered degraded and offers limited ecological value. DH18 is a viable candidate to receive dredged material due to its potential for habitat restoration and its proximity to several state channels in need of maintenance dredging. Within a 5-kilometer radius of DH18, there are 18 state-maintained channels totaling over 20 kilometers in length. The placement of dredged material from local channels is expected to eliminate the stratification of the water column resulting in the reduction of stagnation and hypoxia within DH18. The existing ground surrounding the feature ranges in elevations from approximately -1.8 m to -1.2 m MLW allowing for a high quality shallow water habitat containing beds of submerged aquatic vegetation (SAV) to thrive. Furthermore, the aforementioned CRC water quality samples show that the dissolved oxygen content is significantly greater within the adjacent shallow water.

Figure 8 – Google Earth imagery from 2002 (left) and 2016 (right) clearly depicts the defined depression of Dredged Hole #18

Hydrographic surveys conducted by Gahagan & Bryant Associates in June 2017 show depths as deep as -7 m MLW within the hole and yielded a storage volume of approximately 130,000 cubic meters up to elevation -1.2 m MLW. Dredged material can be placed within DH18 to restore the hole to the surrounding natural elevation between -1.8 m and -1.2 m MLW with an approximate 0.6-m layer of coarse sand to serve as the final cover. Contract documents are currently being developed for a construction project that would result in the dredging of 7 to 9 state channels and placement of approximately 130,000 cubic meters of dredged material into DH18. The construction of this project is anticipated to begin in June 2018. Several measures and best managements practices will take place over the course of construction to ensure minimum adverse effects during dredged material placement. Among these measures are the utilization of a turbidity barrier; monitoring of turbidity, water quality, settlement, compaction, and SAV; and periodic hydrographic surveys.

Given a successful DH18 project, several other CAD site candidates within the New Jersey intracoastal waters are expected to develop into additional dredging construction projects and will result in further restoration of degraded habitat while alleviating the shoaling within New Jersey state maintained channels. The identification of these sites has also greatly enhanced the state-maintained MTS storage capacity potential, and thereby the NJ MTS resilience.

SUSTAINABILITY IN RECOVERY METHODS AND INCREASED RESILIENCE

Based on the example of the NJ MTS, the link between resilience and sustainability is exemplified via economic, operational, and environmental sustainability. Viewing dredged material as a valuable resource, rather than a hindrance, has been critical to envisioning projects that seek to promote more resilient and more sustainable systems through cost effective management of funding sources and use of dredged material to promote marsh health, mitigate coastal erosion, enhance coastal protection and restore habitat.

Economic Sustainability

For many years prior to Superstorm Sandy, funding for maintenance of the Atlantic Shore portion of the NJ MTS was unreliable and not commensurate with maintaining the system in a state of good repair. Following the storm, considerable funds were released to initiate recovery efforts. In addition,

FEMA reimbursed the NJDOT for the cost associated with removing sediment that could be attributed to Superstorm Sandy. The improvement in MTS navigability shown in Figure 2 would not have been possible without the corresponding infusion of funds, and the system's resilience would have been greatly reduced. Compared to Figures 2 and 4 (right), an inverse relationship to funding over time could easily represent the funding received by OMR and its impact on the MTS recovery.

OMR has utilized a number of management tools to increase the efficiency of the design, permitting and procurement of dredging projects. These include bundling channels in close geographic proximity to achieve economy of scale, actively soliciting dredging industry awareness in the Department's efforts through public outreach, and working with resource agencies to increase the number of options available for management of dredged materials, particularly beneficial use. OMR is in the process of developing a management plan for existing CDFs in order to ensure that capacity is maintained for future maintenance needs. Rather than considering CDFs to be for permanent storage, CDFs are considered to be temporary dewatering facilities, and the sites on which they are located to be management facilities for the processing and staging of sediment for beneficial use. As a permanent facility, it is easier to justify the expense of mitigation efforts such as relocation above storm surge elevation, hardening of dikes against waves, or other similar strategies.

OMR is utilizing modern transportation system tools to improve the management of the Atlantic shore MTS. Asset management tools in the form of public facing electronic databases and crowd sourcing of condition reports will enable OMR to better allocate limited funds to the parts of the system with the greatest need. In addition, a strong marketing plan is being implemented for the beneficial use of dredged material that will encourage its use in construction and development. This effort includes a continued partnership with the NJDEP to streamline permitting of projects that utilize dredged material for remediation. Finally, OMR is funding research and pilot projects to evaluate and develop new (or new to NJ) strategies for the beneficial use of dredged material in habitat restoration and recovery efforts. The combination of these initiatives should directly reduce the costs of dredging and dredged material management by increasing competition.

However, none of these efforts will bear fruit without continued state funding at meaningful levels. In fact, the recovery of the NJ MTS will be halted, if not reversed, and MTS resilience dampened if funding levels are dropped back to pre-storm levels. OMR continues to analyze system data to provide the information needed to determine the proper level of funding needed to ensure that this does not happen. At a minimum, the efforts undertaken to date increases the likelihood of a more rapid recovery and a similar anti-fragile response following the next storm

Operational Sustainability

If a CDF is simply used once and filled to capacity, the storage capacity of a MTS would constantly be at a deficit to the system's dredging needs. However, by beneficially using dredged material, the dredged material capacity of the MTS can be continually renewed without having to find, purchase, permit and construct new facilities. Such is the case with the NJ MTS, where dredged material is viewed as a resource rather than a hindrance. As evidenced by the recovery along the Raritan Bayshore region, stored dredged material is often used to elevate existing CDF confining dikes, providing a double benefit of emptying the CDF while also increasing the site's maximum capacity. Rather than only implementing a closure plan for each CDF, increased operational sustainability of the MTS could be achieved through preparing operations and maintenance programs for each site. Coupled with a robust beneficial use program, such as described by Marano, M. et al (2016), such programs would allow for perpetual use of a CDF even after achieving its maximum dike height. In doing so, the potential air capacity of the NJ MTS can be maintained at the improved level indefinitely and allow for enhanced resilience by getting dredges in the water quicker following the next storm.

Likewise, the time between excavation events could be lengthened by reducing the amount of sediment entering the channels. Significant stretches of the NJ coastline are eroding each year and adding to the sediment load within the NJ MTS. Several efforts are underway to reduce dredging needs, including realigning channels to better avoid areas of high sedimentation and making use of natural processes such as tidal scour. The concept of soft armoring via living shorelines and marsh restoration has also been advanced in NJ to reduce erosion rates in the face of sea level rise.

Environmental Sustainability

Several benefits can present themselves when dredged material is used to enhance habitat. Not only does the project facilitate restoration of a navigable waterway but it also promotes a healthier aquatic ecosystem necessary for coastal communities and fishing to thrive. Using dredged material to enhance degraded benthic habitat provides a solution to an age-old problem of dredged material disposal, but because there are a finite number of dredged holes available, this option is not nearly as attractive as marsh restoration.

Around the world, many coastal marshes are disappearing at an alarming rate, including Louisiana, New Jersey, and the U.S. East Coast in general as shown by Barras, J.A. (2006) and Leatherman, S.P. et al. (2000). Combined with subsidence, sea level rise unbalances the marshes health by shifting the intertidal zone and drowning marsh vegetation. The resulting loss of vegetation reduces the associated root mat and contributes to erosion of these critical coastal ecosystems, often depositing the eroded material into navigable channels and increasing the need to dredge. As such, the stabilization of these marshes via thin-layer placement (TLP) of dredged material would not only restore critical habitat; but, in doing so, could potentially reduce the need for maintenance dredging in adjacent channels. Application of TLP to degraded marshes could be performed on a cyclical basis to keep up with sea level rise while fulfilling the dredged material storage needs of a channel maintenance program. Additional benefit is provided via increased resilience to coastal communities behind these coastal marshes, as a marsh buffer has been shown by Lunemann, M. et al. (2017) to reduce damage beginning at buffer widths of 45 meters. Building on two pilot projects by the US Army Corps of Engineers at Stone Harbor, NJ and Avalon, NJ, OMR sought to expand on these examples with a joint pilot project for TLP using dredged material to enhance marsh health at Fortescue, NJ.

Case Study: Thin-Layer Placement – Fortescue, New Jersey

Between the winter of 2016 and spring of 2017, maintenance dredging was performed in Fortescue Creek located in Downe Township, Cumberland County, New Jersey. The project consisted of over 27,500 cubic meters of material that was dredged and placed at 4 different sites. Fill methods consisted of thin-layer placement as marsh fill, non-bathing beach fill, shoreline restoration dune fill, and bathing beach fill. Material was designated to a placement site after extensive material sampling and analysis of the shoal volumes within Fortescue Creek. Portions of the channel containing the most fine-grained sediment (silts and clays) were designated most suitable for marsh fill. Portions of the channel consisting of coarse grained material greater than 90% sand were designated for either dune fill or bathing beach fill, while material containing between 75% and 90% of sand content was designated for non-bathing beach fill. Aside from being unique for utilizing four separate placement areas, this project was a pilot test of thin layer marsh enhancement sponsored by the New Jersey Department of Environmental (NJDEP) on 8 hectares of degraded coastal marsh. Collaboration between multiple entities, including NJDOT-OMR, USACE, and NJDEP was required during the planning phase to establish goals and success criteria. Extensive studies of the marsh were performed by the NJDEP's environmental consultant to determine a range of target fill elevations that would restore areas of low and high marsh habitat without creating new uplands. Upland habitat in marsh areas of this part of New Jersey are prone to domination by *Phragmites australis*, an invasive weed with little habitat value.

As with any pilot project, challenges were presented throughout both the design and construction processes. Given the geography of the project location, the adjacent Delaware Bay plays a major role in dictating tidal fluctuation at this site. Tidal ranges varied from approximately -0.152 m to +2.124 m MLW in the most severe instances. One challenge associated with a large tidal range at a degraded marsh site is that portions of the marsh fill area can become completely inundated during the peak of a high tide event. For this reason, specialized equipment was utilized to transport materials and personnel across the work site. Equipment used during the marsh fill events included pontoon excavators, marsh buggies, an amphibious tracked vehicle, and a low ground pressure bulldozer. Dredging was performed using a 0.3-m discharge pipe diameter cutter suction dredge and material was conveyed through a 0.3-m diameter high density-polyethylene (HDPE) pipeline. The pipeline from the stern of the dredge was landed ashore of the marsh and connected to a network of additional shore pipeline featuring a "trunk" with several "branches" that spanned across the entire marsh fill area. Each branch had the ability to be opened and closed by means of a valve system at its intersection with the trunk. This methodology was chosen for a variety of reasons, such as allowing for a rapid response to divert the flow in the event of insufficient control of the inflow point, and to allow for simultaneous inflow at multiple locations. The use of HDPE pipeline allows for additional maneuverability of the inflow points given its flexible and lightweight nature as indicated in Figure 9.

Figure 9 – Pumping Dredged Material onto the Marsh during Fortescue TLP

To further control inflow material, multiple layers of sediment control/filtration (SiltSoxx), were placed along the perimeter of the placement site as shown in Figure 10. The SiltSoxx played an important role in capturing fine-grained dredged material that failed to separate from the slurry upon discharge, which prevented sediment from reaching natural drainage features outside the placement area and being reintroduced into the surrounding waterways.

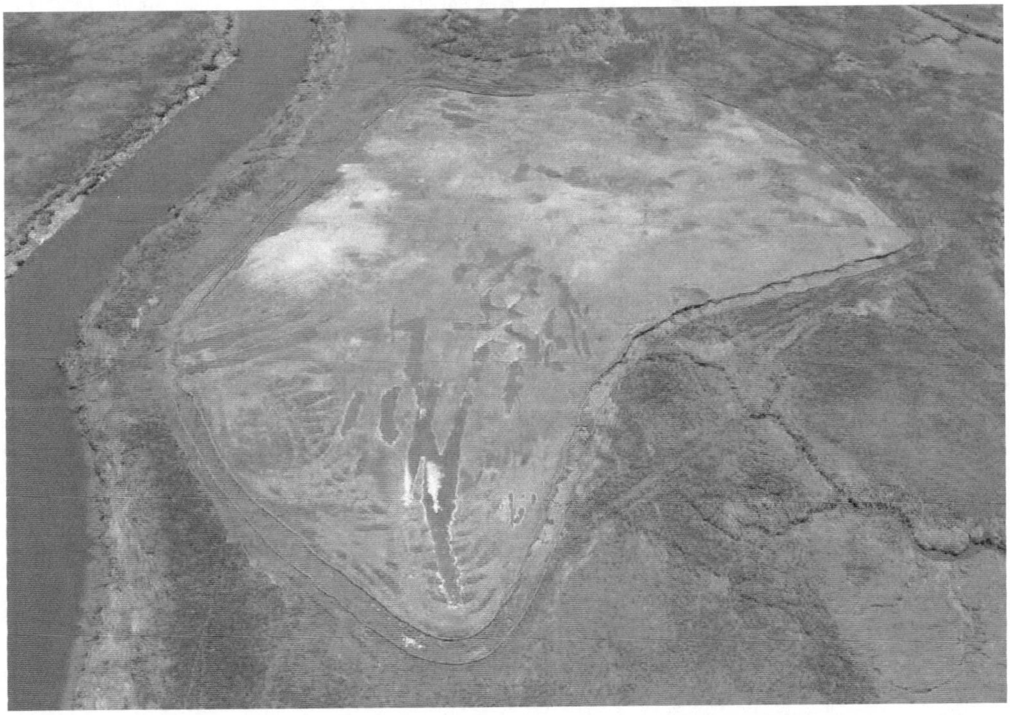

Figure 10 – Aerial View of Fortescue TLP Site and Outer Rows of Sediment Control

After all thin-layer placement operations had been completed; approximately 5,350 cubic meters was placed across the site. As of January 2018, after one full growing season, the marsh had already established desirable vegetation and was on its way to environmental success.

Over the course of placement operations, the design team discussed several ways to improve thin-layer placement methods going forward. One major challenge associated with this concept is maximum fill elevation restrictions. In the event that the target fill elevation is exceeded, the potential for invasive plant species, such as *Phragmites australis* or Common Reed, increases. In cases where a single inflow point is used to discharge for an excessive amount of time, the inflow location is susceptible to accumulate too much material and form mounds above the target elevations. Appropriating the amount of time spent at a single discharge point is dependent on the thickness of fill, characteristics of the dredged material, and volume of slurry. In order to combat mounding, either the discharge point must be moved frequently, or the slurry inflow rate must be reduced. This was experimented with in Fortescue by utilizing amphibious excavators and marsh buggies to physically adjust the discharge point in a lateral manner, moving the pipeline from side to side to encourage dredged material to disperse.

Another method associated with thin-layer placement is referred to as spray dredging. Spray dredging is conducted by discharging the slurry through a narrow opening, such as a nozzle, with the end of pipeline oriented in an upward fashion. This results in an increase in discharge velocity and therefore will spray dredged material allowing for a more controlled filling over a greater coverage area. A disadvantage to this technique is that additional risks are associated with transportation of heavy equipment across sensitive marsh grasses, as was experienced during the Fortescue project. Excessive movement of the equipment on soft soils, especially excavators mounted on pontoon tracks, would occasionally lead to bogging. Furthermore, reducing the discharge diameter by use of a nozzle can result in losses of dredge productivity and a greater propensity for clogging, leading to increased project costs and longer project durations.

Another notable conclusion taken from the Fortescue project was that the high tide events served to further distribute dredged material in a uniform manner. These tides combined with wind and precipitation were beneficial in achieving more acreage of fill as well as reducing the height of undesired mounds. Further consideration should be given to the use of natural processes to "smooth out" the final elevations in future TLP projects.

Given the vast acreage of similar ecosystems along New Jersey's coastline, there are many additional opportunities for thin-layer placement. Applying innovative strategies built upon lessons learned from pilot projects such as Fortescue is an important part of developing a more sustainable and resilient process.

Mutual Improvement to Sustainability and Resilience via TLP

The approaches described by the Keansburg CDF (i.e. CDFs used for coastal protection) and Fortescue TLP (i.e. dredged material for enhancing marsh health) provide an unorthodox approach to managing dredged material in New Jersey, but one that seeks to achieve a win-win situation out of singular projects and promote public perception of dredged material as a resource. The potential MTS resilience gained from the sustainable TLP approach is especially promising.

The application of the TLP approach to the coastal marshes around the state's upland CDFs would provide enhanced protection to the confining berms as noted by Lunemann, M. et al. (2017), thereby boosting MTS resilience while maintaining the health of the critical marsh habitat. However, New Jersey contains a significant amount of coastal marshes along the eastern back bays. Using TLP on these marshes would provide dredged material as a sustainable resource to promote marsh health in the face of sea level rise along roughly 145 kilometers of New Jersey's coastline from Bay Head to Cape May, while also providing valuable dredged material storage capacity on a regular basis.

Looking at the Barnegat Bay region alone (Figure 11), Barone, D.A. et al. (2014) have identified TLP as an alternative at three locations to restore wetlands eroded from Superstorm Sandy directly seaward of structures, which would require nearly 145,000 cubic meters of dredged material. On a wider scale, based on the Barnegat Bay Partnership's 2016 State of the Bay Report by Baldwin-Brown, A., et al. (2016), TLP could be used to restore roughly 2.75 square kilometers of tidal marsh lost to erosion since 1995. Assuming an average of only 0.15 to 0.3 meters placed to restore these eroded wetlands, a potential dredged material storage capacity of approximately 400,000 to 800,000 cubic meters could be gained.

While these potential applications deal with restoring already-lost tidal marshes, degraded marshland could also benefit from TLP of dredged material. Based on mapping of confirmed stressed marshland within the Barnegat Bay by Baldwin-Brown, A. et al. (2016), these stressed marshland areas total approximately 19 square kilometers. By using TLP to restore this marshland, an estimated additional

1.45 million cubic meters of storage capacity could be gained, assuming TLP at a conservatively low average application thickness of 7.5 centimeters for the magenta-shaded areas of coastal marsh in Figure 11. Only areas of marsh confirmed as stressed via sampling and testing are provided in Figure 11, and additional areas of degraded marsh are likely present.

Figure 11 – Areas of Stressed Marsh in Barnegat Bay Region (shaded magenta), with Channels (in red, left) and without Channels (right)

As a result, TLP use in just the Barnegat Bay region could provide as much as 2 million cubic meters of dredged material storage capacity, effectively quadrupling dredged material storage capacity in this region. Providing a beneficial use alternative in the northern Barnegat Bay where no CDFs exist, and promoting a healthy marsh habitat and coastal buffer along 65 kilometers of coastline, enhances both sustainability of the region and resilience of the MTS. The availability of these degraded marshes for dredged material storage could significantly increase the rate of recovery systemwide following the next storm and shift the channel navigability to a higher level of resilience.

The state is currently in discussions with the US Fish and Wildlife Service to restore numerous degraded coastal marshes in Barnegat Bay via TLP using material dredged from the NJ MTS. Considering the success of the Fortescue TLP project and that these marshlands are all within an 8-kilometer radius of state-maintained channels, the potential for TLP to play an increased role in the state channel dredging program is high.

However, while the agency mindset toward dredged material use and TLP may be optimistic in New Jersey, the current cost of TLP (US$150/cubic meter) is approximately double that of CDF-use, making TLP less attractive economically. Increased frequency of these projects, as well as increasing the scope and scale of TLP projects, is hoped to result in reduced costs or more innovative methods that would make TLP more economically sustainable. One option to recoup TLP costs could be the formation of a state-owned wetlands bank. Under this setup, the state would gain regulatory credits for restoring wetlands using TLP that could then be sold to private developers, or used by NJDOT, to offset impacts to wetlands on other projects. Depending on location, wetlands bank credits have been sold for as high as US$150,000 per credit in New Jersey with an average in-state value of approximately US$50,000 per credit based on the NJ In-Lieu Fee Mitigation Program, Final

Instrument (2015). While not likely to fully fund NJ MTS projects, the value of these credits could seek to offset some of the costs associated with TLP and make TLP more economically sustainable.

CONCLUSIONS

Superstorm Sandy has provided a daunting recovery challenge to the region that has sparked discussions on resilience and sustainability in the storm's aftermath. Numerous challenges faced the recovery process, including extensive debris removal, severe shoaling of navigation channels, and a change in channel management agencies. Though these challenges dampened the resilience of the NJ MTS following the storm, OMR's prioritization and pre-recovery planning contributed to higher resilience on the regional level, as demonstrated by the Raritan Bayshore region. OMR's efforts to restore the state-maintained marine transportation system, as well as their commitment toward enhancing sustainability and resilience as part of their recovery efforts to reduce future damage and improve dredged material storage capacity, have yielded an anti-fragile response as indicated by Figures 2, 4 (right), & 5. As such, OMR has demonstrated that resilience planning, even if performed during the initial stages of recovery, provides a more effective and resilient recovery which can lead to anti-fragility of a system.

A direct link between sustainability and resilience was also highlighted by the NJ MTS following Superstorm Sandy. With post-storm funding levels significantly exceeding pre-storm levels, the funding stream made available following the storm also contributed to the anti-fragile response. However, failure to maintain a sustainable, adequate funding source would result in undoing or reversing the gains in resilience and hamper future recovery efforts as a direct result of the lower economic sustainability of the system.

However, in addition to economic sustainability, a more resilient MTS requires a comprehensive approach to dredged material management to maintain dredged material disposal capacity. By viewing CDFs as temporary dewatering facilities, using CAD to improve degraded aquatic habitat, and by applying TLP of dredged material to enhance degraded coastal marshes, win-win situations are developed via gains in both operational and environmental sustainability. The success of the Fortescue TLP project demonstrates the efficacy, challenges, and potential of this approach to enhance NJ MTS resilience on a wider scale. The increased potential storage capacity directly leads to improved resilience, as dredging projects could begin more rapidly following the next event, while additional economic sustainability could be gained by establishing wetlands banks to partially offset the costs of TLP projects.

The importance of multi-party coordination and stakeholder engagement during recovery processes was indicated by the Keansburg CDF case study. Via stakeholder engagement, a unique situation of locals wanting more dredged material opened the door to the creation of additional storage capacity by creating a hybrid CDF-dune system to enhance coastal protection through the beneficial use of dredged material. The debris removal phase of the NJ MTS recovery provided further correlation between multi-party coordination and the effectiveness and speed of recovery.

As such, with a little foresight and commitment to resilience, extreme events or other stressors can result in an improved system compared to the pre-stressor conditions, i.e. anti-fragility, due to a refocus of political priorities, availability of improved funding, and changes in public perception. As the OMR program continues to advance and grow, the resilience of the NJ MTS grows as well. The continued improvement of the state-maintained channels to a state of good repair and continued efforts to increase available dredged material storage capacity, will find the NJ MTS in a better condition when the next storm impacts the state and in a position to better resist damage and more rapidly recover.

ACKNOWLEDGEMENTS

The authors would like to recognize the following people for their involvement and contribution to the work that formed the basis of this paper: Sean Flanigan & John Drake from Gahagan & Bryant Associates (GBA); Jerry Chen, Steve Chiaramonte, Kristen Maloney, & Dave Wysocki from WSP; Kelly Rosofsky from Dewberry Engineers; and Bill Henderson & Joselyn Wall from NJDOT OMR.

REFERENCES

Baldwin-Brown, A., Baker, R., Caruso, C., Gaylord, B., Karl, R., Ruocco, W., Schuster, B., Spinweber, B., Vasslides, J., and Wieben, C. (2016). "State of the Bay Report | 2016", Barnegat Bay Partnership, Toms River, New Jersey.

Barone, D.A., Howard, B.S., Ferencz, A., McKenna, K., Macdonald, A., and Orgera, R. (2014). "Beneficial Use of Dredged Material to Restore Wetlands for Coastal Flood Mitigation, Barnegat Bay, New Jersey", *Report for The State of New Jersey, Governor's Office of Rebuilding and Recovery*, Trenton, New Jersey.

Barras, J.A. (2006). "Land area change in coastal Louisiana after the 2005 hurricanes—a series of three maps" *U.S. Geological Survey Open-File Report 06-1274*.

Leatherman, S.P., Zhang, K., & Douglas, B.C. (2000). "Sea Level Rise Shown to Drive Coastal Erosion", *Eos, Transactions American Geophysical Union*, Vol. 81, No. 6, p. 55-57.

Linkov, I., Bridges, T., Creutzig, F., Decker, J., Fox-Lent, C., Kroeger, W., Lambert, J.H., Levermann, A., Montreuil, B., Nathwani, J., Nyer, R., Renn, O., Scharte, B., Scheffler, A., Schreurs, M., and Thiel-Clemen, T. (2014). "Changing the resilience paradigm", *Nature Climate Change*, Vol. 4, No. 6, Macmillan Publishers Ltd., London, p. 407-409.

Lunemann, M., Marano, M., and Douglas, W.S. (2017). "Resilience of Upland Confined Disposal Facilities and Beneficial Re-Use of Dredged Material for Coastal Protection", *Proceedings of The Dredging Summit & Expo '17*, Western Dredging Association, Vancouver, BC, p. 144-157.

Marano, M., Douglas, W.S., and Clifton, G. (2017). "Rebuilding "Stronger than the Storm" – Enhancing Resilience of the New Jersey Marine Transportation System", *Proceedings of The Dredging Summit & Expo '17*, Western Dredging Association, Vancouver, BC, p. 272-286.

PIANC (*2018 pending*) "Resilience of the Maritime and Inland Waterborne Transport System", Final Report of Task Group 193, PIANC, Brussels.

"State Channel Maintenance Capacity" (2015). New Jersey Department of Transportation's Office of Maritime Resources, Trenton, New Jersey.

"State of New Jersey In-Lieu Fee Mitigation Program, Final Instrument" (2015). New Jersey Department of Environmental Protection, Trenton, New Jersey.

Wakeman, T.H. & Miller, J. (2013). "Lessons from Hurricane Sandy for Port Resilience", University Transportation Research Center – Region 2, The City College of New York, New York, NY.

Wakeman, T.H., Miller, J., and Python, G. (2015). "Port Resilience: Overcoming Threats to Maritime Infrastructure and Operations from Climate Change", University Transportation Research Center – Region 2, The City College of New York, New York, NY.

TOWARDS IMPROVED PREDICTION OF DREDGING PLUMES: NUMERICAL AND PHYSICAL MODELLING

by

Boudewijn Decrop[1] and Marc Sas[2]

ABSTRACT

Construction and maintenance of ports and waterways involves dredging activities in many cases. Dredging projects require assessment and mitigation of a number of environmental impacts. Some of the potential impacts are related to turbidity plumes resulting from hydraulic and mechanical dredging processes bringing sediment into suspension.

In the recent past, environmental awareness and by consequence environmental legislation has become stronger. As a result, dredging contractors and dredging consultancy have been faced with the challenge to implement better control mechanisms for environmental management purposes. More specifically, turbidity plumes have been monitored closely in the past to follow up on their fate. Numerical simulations allow for real-time forecasting of the fate of the turbidity plumes in the near future. By means of a well-calibrated tidal flow model, planned dredging activities can be implemented as sediment sources in the numerical flow models. In this way, the plume dispersion due to interaction of the tidal flows and the timing of activities spilling sediments can be predicted up to a week ahead.

In the past, large-scale numerical flow models have been applied, and covered the wider areas around the project site that can potentially be affected by the works. Overflow losses from Trailer Suction Hopper Dredgers (TSHD) are one of the main sediment spills during the execution of dredging projects. In the past, near-field sediment distributions from overflow spills have been determined using simplified laws and crude estimates of losses.

In the presented work, efforts have been made to improve the accuracy of plume simulations by performing highly-detailed computational fluid dynamics (CFD) simulations of the flows of the water-sediment-air mixture around the ship hull and its interaction with the propellers. These detailed simulations have several benefits, such as assessment of overflow design and insights in the three-dimensional distribution of sediments near the dredger, but are too time-consuming to be used in operational forecasting of turbidity plumes. In the work presented in this paper, the CFD results have been applied to develop a parameterized model, significantly faster compared to the CFD simulations, but more accurate compared to the previous generation of near-field models.

The coupling of this new generation of near-field spill models with the far-field (large scale) flow models allows for a significant increase in accuracy of turbidity forecasting. In this way, using forecasting models, dredge and disposal productions can be optimised while complying with turbidity levels imposed in the environmental criteria.

The presented paper has been submitted before to the jury of the De Paepe-Willems Award 2017 and went on to receive the first prize, handed over to the first author at the PIANC AGA in Cairns, Australia on June 19th, 2017.

[1] International Marine and Dredging Consultants (IMDC), Belgium, Boudewijn.Decrop@imdc.be

[2] International Marine and Dredging Consultants (IMDC), Belgium, Marc.Sas@imdc.be

1. INTRODUCTION

Construction and maintenance of ports and waterways involves dredging activities in many cases. Dredging projects require assessment and mitigation of a number of environmental impacts. One of these impacts is increased turbidity and sedimentation due to dredging and disposal activities bringing sediments in suspension. Reduction of light penetration, increased sedimentation and increased suspended sediment concentrations can have potentially adverse effects on sensitive habitats (eg. coral & seagrasss ecosystems) or nearby human activities (eg. aquaculture, industrial/drinking water intakes). The extent of the impacts will depend on the quantity, frequency and duration of dredging, adopted methodology, site-specific conditions (wind, wave and current fields, grain-size distribution and water depth), proximity to sensitive sites and tolerance of living organisms to altered turbidity conditions (PIANC, 2010).

Increased awareness has instigated stricter environmental legislation related to these activities. Project environmental permits often stipulate project-specific regulations, which can entail strict turbidity thresholds for these activities. Operational turbidity management in these projects is warranted, as exceeding turbidity thresholds can trigger corrective measures, increased monitoring efforts, relocation of dredge activity, a decrease in or –worst case - a cease of dredging and dredge spoil placement activities.

The modelling tools presented in this paper add to the development of a system in which real-time predictions of the plume behaviour can be achieved. The operational planning of dredge operations for the next few days can be implemented in a forecast model environment, In case a breach of turbidity thresholds is predicted, the operational planning can be revised or altered to avoid breaches.

Trailer Suction Hopper Dredgers (TSHD's) often deploy an overflow system through which excess sea water is skimmed from the hopper (Figure 1). The released water contains a varying concentration of fine sediment material, of which a fraction can form turbidity plumes. In the presence of currents, these plumes are advected over longer distances. The plumes can affect environmentally sensitive areas throughout coastal, river or offshore systems (Bray, 2008).

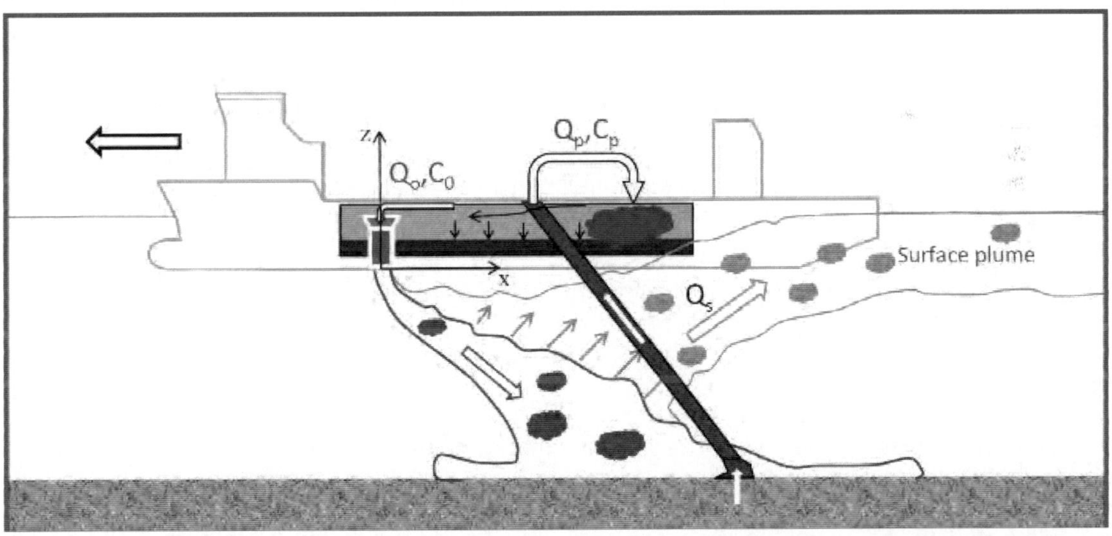

Figure 1. Sketch of a Trailer Suction Hopper Dredger, pumping a water-sediment mixture (mass concentration C_p) towards the hopper, at volume discharge Q_p. The sediment mixture settles while flowing towards the overflow, where a lighter mixture with sediment concentration C0 is released, with discharge Q_0.

Today, these environmental impacts are assessed using predictive simulations based on large-scale modelling of dredging scenarios. Since large-scale modelling of (tidal) currents in estuaries and regional seas is usually executed using a simplified, hydrostatic form of the Navier-Stokes equations (Saint-Venant equations), it is useful to divide dredging plumes in two parts. The first part is the near-field section of the plume, near the exit of the overflow shaft. This part is called the dynamic plume since it is still under influence of the ship and the excess density of the water-sediment-air mixture. The excess density of the plume can be expressed as

$$\Delta\rho=\rho_m-\rho_\infty=c(1-\rho_\infty/\rho_s)-\phi_a(\rho_\infty-\rho_a) \quad (1)$$

where ρ_∞ is the mass density of the sea water, ρ_m the mass density of the sediment-water mixture in the plume, c is the sediment mass concentration, ρ_s the mass density of the sediment material, ϕ_a is the volume fraction of air bubbles and ρ_a is the mass concentration of air bubbles.

The second part of the plume starts where the released mixture is diluted to the point where the plume bulk density ρ_m is no longer significantly higher compared to the surrounding sea water. This part is called the passive plume since it is passively advected with currents in the sea waters, with settling of flocs and benthic aggregates as main process (e.g. Smith and Friedrichs, 2011). Large-scale modelling tools are capable to solve the passive part of the plume. However, in the modelling of dredging plumes the dynamic part of the plume is still a missing link between the sediment discharge at the overflow exit and the passive part of the plume. Today this gap is bridged by roughly estimating the sediment flux to the passive plumes, as a fixed percentage of fines in the production. It has been shown that this percentage can vary widely, even within one loading cycle (Decrop *et al.*, 2014; de Wit *et al.*, 2014).

In the past, the near-field modelling done for the transformation of the bulk overflow outflow to a vertical distribution of the sediment behind the vessel has also been executed with integral models representing the integrated Navier-Stokes equations for a buoyant jet in crossflow (Fischer, 1979; Jirka, 2006). Such models were implemented by e.g. Spearman (2011). However, this type of model cannot incorporate the formation of a surface plume due to the complex flow pattern around the vessel and due to air bubbles. Instead, a mere constant factor can be applied.

In this paper, an overview is given of the tools developed by the author in the recent past, more specifically a highly detailed 3D CFD model and a fast parameter model. Today's application of the developed tools is discussed, leading to improvement of turbidity assessment in planning phase and in operational phase. The presented research was executed by IMDC with additional funding from IWT in Belgium. Special thanks go to professors De Mulder (Hydraulics Laboratory, Ghent University) and Toorman (Hydraulics Laboratory, KULeuven) for supervising the research.

2. NEAR-FIELD CFD MODEL

A 3D numerical simulation model has been developed in the Ansys Fluent™ environment. The aim of this Computational Fluid Dynamics (CFD) model is to represent accurately the flow patterns of the water-sediment-air mixture in the direct vicinity of a hopper dredger while trailing.

In the following sections, a short overview of the model development is given, as well as results of the analysis of influencing factors on plume dispersion.

2.1 Set-up and validation

The 3D CFD model solves for the velocity vectors of three phases: water, sediments and air bubbles. The following approach was chosen to handle the three phases. First, the momentum and continuity equations for a mixture of water and sediments are solved. Then, the relative velocity (slip velocity) of the sediment compared to the water velocity is determined by including the effects of settling and drag (Manninen *et al.*, 1996) and including the effect of turbulent diffusion using a drift flux term. This method is allowed when the expected slip velocity is low. The slip velocity of air bubbles in water has a much higher range compared to the sediments in the overflow plume, which are mainly fines. Therefore, a different approach was used for the air bubble dispersion. It was solved using a Lagrangian approach in which the acceleration vector is determined from a force balance consisting of drag, gravity, virtual mass and pressure gradient (Decrop *et al.*, 2014).

The input of momentum and swirl due to the propellers is modelled using an actuator disk approach. In this approach, the axial and tangential velocity components - as a function of the radial distance from the hub - are imposed by implementing a pressure jump over a circular surface in the model grid.

The actual geometry of an existing TSHD is embedded in the model grid. The model domain is discretised using an unstructured grid. This allows for an accurate representation of the complex geometry of the ship. Refined grid cell layers are included to resolve the velocity profiles at the wall boundary layers at ship hull and in the overflow shaft (Figure 2). Additional grid refinements were included in regions of plume presence, strong gradients, strain near the bow, propeller flow and near the sea bed. The surface mesh at the hull was refined in zones of strong curvature to represent the shape in an optimal way.

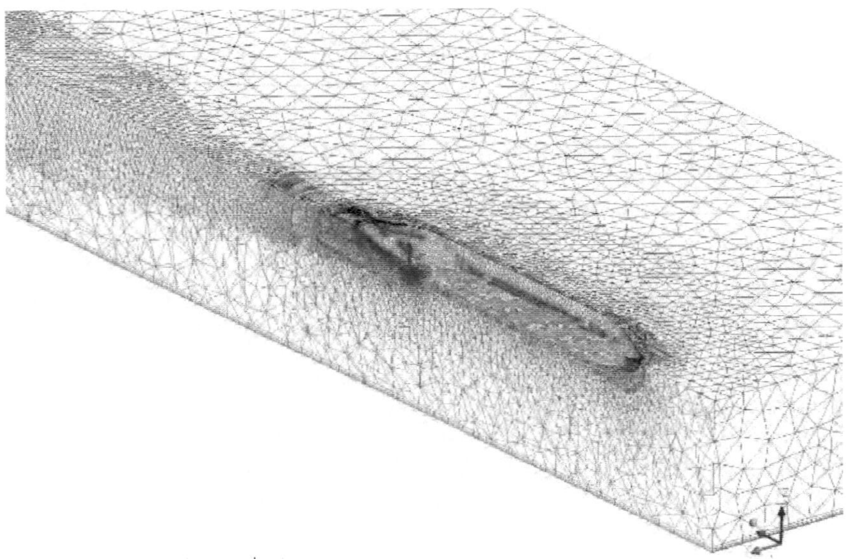

Figure 2. Example of a model grid, tailored to a specific plume case for computational efficiency. Grid is sliced along the ship axis. Ship hull grid in brown, water surface grid in black, subsurface grid along slice in blue.

The turbulent flow field was solved using the Large-Eddy Simulation (LES) technique, in which the larger turbulent vortices are explicitly resolved on the grid. Amongst other reasons, this is needed to include the interactions between an individual vortex and the local sediment gradients near the edges of the plume.

The CFD model was validated in a number of steps. Two major steps were taken: validation of a laboratory-scale model (Decrop *et al.*, 2015a) and validation of a full-scale model. A thorough validation of a laboratory-scale CFD model was executed based on measurements taken in a physical model. The CFD model results were compared to highly-detailed measurements of sediment concentration, turbulent sediment fluxes, mean flow velocity components (*U, V, W)* turbulent velocity fluctuations and, finally, the Reynolds stress

After the detailed validation at laboratory scale, the CFD model was converted to a full-scale model, including a realistic vessel geometry, propellers and overflow shaft (Figure 3).

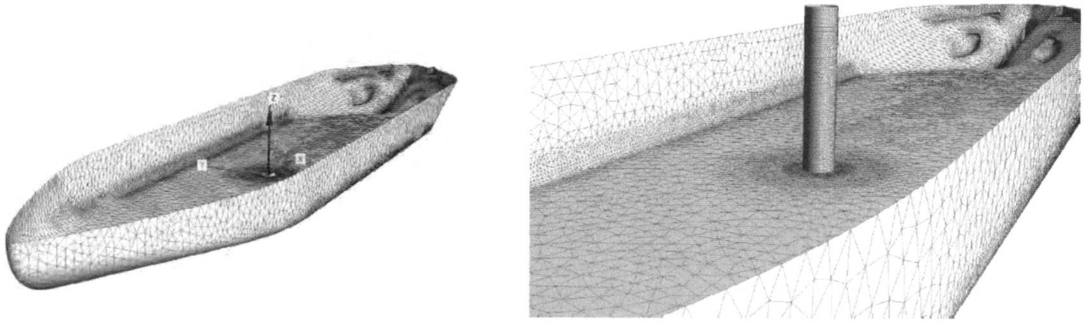

Figure 3. Details of the surface mesh at the TSHD hull and overflow shaft walls.

A measurement campaign was set up to measure a real-life overflow dredging plume in the field (Decrop and Sas, 2014). Detailed vertical profiles of sediment concentration were recorded throughout the full water column, from surface to 2 cm above the sea bed. In this way, also the near-bed highly-concentrated mud layers could be monitored. Also, sediment concentration observations were conducted closer to the surface and along the complete plume using Optical Backscatter instruments towed behind the survey boat. Finally, Acoustic Doppler Current Profiler (ADCP) measurements have been taking, allowing for the visualisation of the concentration levels along a vertical slice of the plume.

Subsequently, CFD simulations were set up representing identical ambient conditions as observed in the field. The results of the simulations were compared with the field data, as a validation exercise for the full-scale CFD model. Suspended sediment concentration measurements in the surface plume matched well with the CFD model (Figure 4, upper panel). Also, observations of the deeper parts of the overflow plumes corresponded well with the CFD model (Figure 4, lower panel).

Two examples of results of the CFD model are shown in Figure 5. In this example, the head-current is relatively strong, leading to a significant surface plume (with the potential to travel long distances).

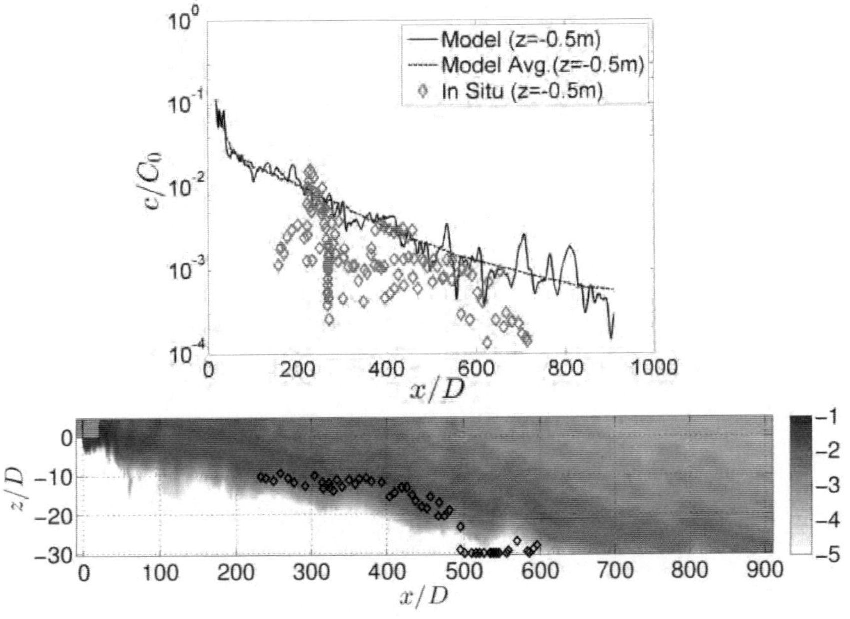

Figure 4. Upper panel: range of observed sediment concentration c/C_0 (red markers) compared to the centerline (maximum) c/C_0 values from the CFD model. Lower panel: Vertical slice of CFD sediment concentration $\log(c/C_0)$ (along the ship axis, in grayscale), compared to the lower edge of the plume as observed in the field (black diamonds).

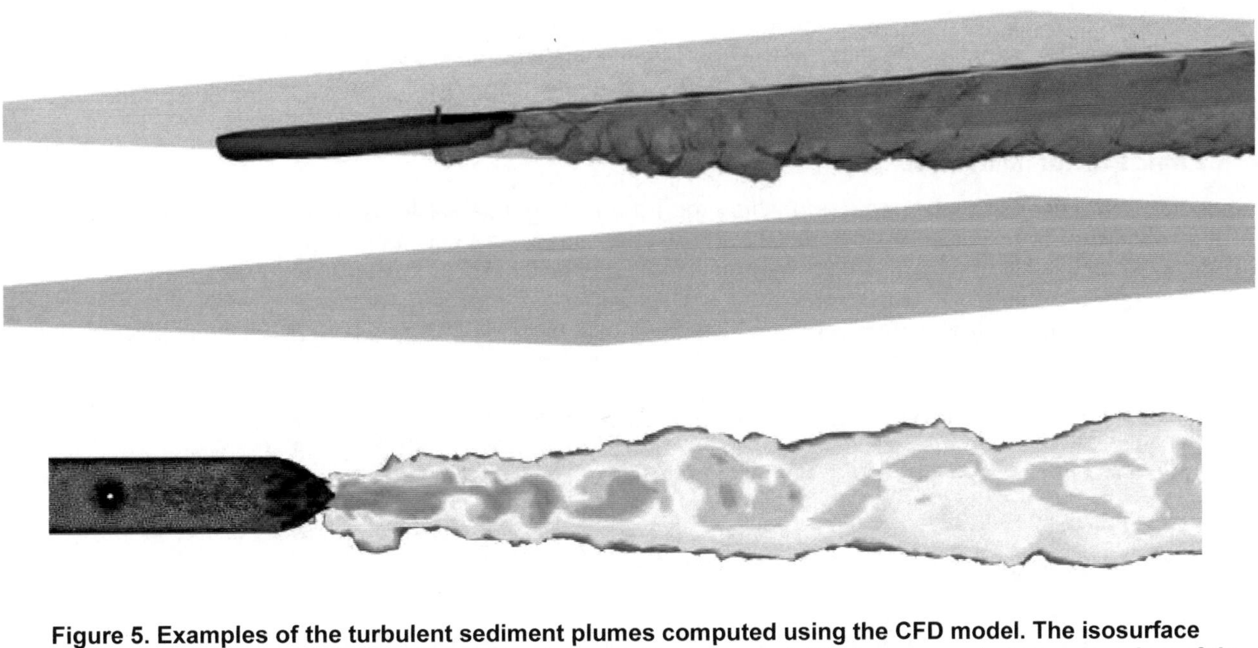

Figure 5. Examples of the turbulent sediment plumes computed using the CFD model. The isosurface connecting the locations at which $c/C_0=10^{-4}$ is shown in the top panel. The lower panel gives a top view of the computed sediment concentration at the surface.

2.2 Sensitivity of plume dispersion to boundary conditions

The main goal of near-field CFD modelling is to compute the vertical and horizontal distributions of the overflow behind the ship, for far-field model input. However, the full three-dimensional fields of flow velocity and concentrations of sediment and air bubbles are also available for analysis. For example, the influence of a number of operational and environmental parameters has been studied by comparing two simulations in which only one parameter has been changed. These insights have in later stages been used for the development of simplified models.

2.2.1 Influence of dredging speed

First, two simulations are carried out in which a water-sediment mixture is released with a varying average velocity (Figure 6). It is shown that the stronger relative flow velocity induces an increase of the surface plume sediment concentration with a factor 10.

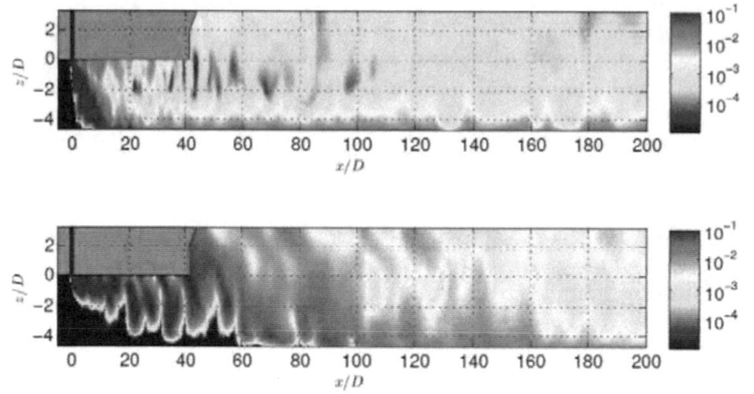

Figure 6. CFD results (relative concentration c/C_0) with low dredging speed (top panel) and with high dredging speed and/or head current (lower panel).

2.2.2 Influence of mixture density

Secondly, simulations are carried out in which the flow velocity relative to the ship, *U0*, is kept constant at 1 m/s, while the other parameters are equal to previous case. The overflow sediment concentration C_0 is equal to 10 g/l in one simulation (Figure 7, top panel) and to 150 g/l in the other (Figure 7, lower panel). It can be seen very clearly that the fraction of the released sediments going to a surface plume (surface value of c/C_0) is much higher (up to factor 100) when the overflow mixture is light. The mixture does not have sufficient excess density to descend to the sea bed.

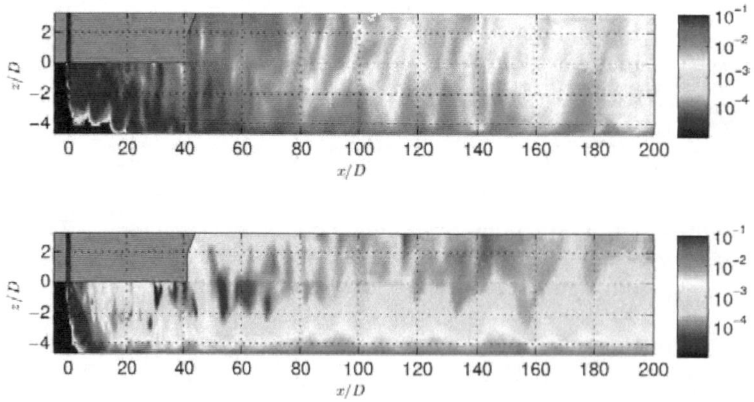

Figure 7. CFD results (relative concentration c/C_0) with low C_0 (10 g/l; top panel) and with high C_0 (150 g/l; lower panel).

2.2.3 Influence of air bubbles

The influence of the entrainment of air bubbles into the overflow shaft was also investigated. More specifically, the influence of the so-called green valve was investigated. The green valve is designed to reduce the turbidity in the water column by choking the flow, reducing the number of air bubbles in the overflow and by consequence reducing the uplifting effect thereof. It is usually assumed that the green valve has an effect under all circumstances. It was shown using the CFD model that the effectiveness of the green valve is largely dependent on the ambient conditions and overflow mixture properties. See Decrop *et al.* (2015b).

3. NEAR-FIELD PARAMETER MODEL

3.1 Introduction

For the study of the behaviour of specific plume cases, or for gaining insights in the effects of operational aspects on the plume behaviour, a CFD model is very valuable. In some phases of a dredging project, however, the long simulation times associated with it are not always acceptable. In the operational project phase, real-time plume predictions are needed to assess the timing and location of dredging in the day-by-day planning of works. At this stage, the long simulation times of the CFD model are prohibitive.

The large-scale simulation of the far-field plumes is generally executed with a shallow-water equations-based hydrodynamic flow model with a sediment transport equation and a source term for the overflow releases. The source term which has to be supplied to the large-scale model needs a vertical distribution. A parameterised model has been designed to perform this task. Essentially, this model is a trade-off between calculation speed and accuracy. It is less accurate compared to a CFD model, but much faster and therefore applicable in cases where the CFD model is not possible, e.g. real-time forecasting simulations.

3.2 Model Set-up

First, the different length scales and fluxes need to be condensed into non-dimensional numbers. This makes the parameterisation of the vertical flux profiles more generic.

In Figure 8, the different scales are sketched. The water depth H is the sum of the TSHD draft H_d and keel clearance H_k. The distance between the overflow and the stern is denoted as L_o.

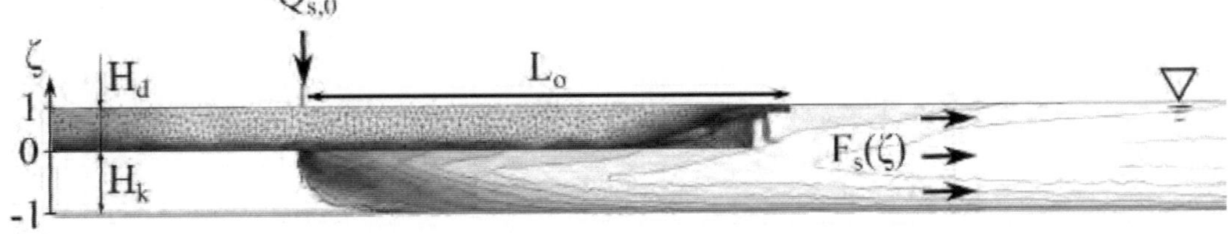

Figure 8. Definition sketch of the parameter model and variables used therein.

The vertical coordinate z can now be scaled to a dimensionless coordinate ζ, equal to -1 at the sea bed, to 0 at the keel and to 1 at the water surface. This transformation allows to make use of Chebychev polynomials for the parameterisation of the shape of the vertical profile of sediment flux.

From a large number of CFD simulations, representing the full range of realistic boundary conditions, results are extracted. These results are used to determine vertical profiles of sediment flux, used as data set to fit the parameters in the parameter model. The profiles are determined as follows. The time-averaged sediment flux *f* in the sediment plume (in kg/(s.m)) is defined:

$$f(x,y,\zeta) = C(x,y,\zeta)U(x,y,\zeta) \tag{2}$$

with C and U the time-averaged sediment concentration and flow velocity.

The flux q_s is integrated over the width of the plume, determined as:

$$q_s = \int_{-B/2}^{B/2} f(x,y,\zeta)dy \tag{3}$$

Where *B* is the width of the plume. At this point we have a sediment flux in kg/s in the plume at every location along x and per dimensionless unit of height (ζ is non-dimensional). A distance x_p needs to be defined at which the vertical profile of q_s is evaluated. A fixed distance is defined at which the CFD model output is evaluated and by consequence at which the parameter model is valid. The distance x_p was chosen at $2.5L_s$, with L_s the vessel length. At this distance from the vessel, the parameter model output is valid for implementation in a far-field model.

The vertical profile of the flux that will be parameterised is non-dimensionalised and defined by:

$$F_s(\zeta) = q_s(x_p,\zeta)/Q_{s,0} \tag{4}$$

where $Q_{s,0}=C_0Q_0$ is the sediment outflow from the overflow, C_0 is the overflow sediment mass concentration, Q_0 is the volume discharge through the overflow.

For each CFD result in the data set, the profile $F_s(\zeta)$ is determined at $x_p=2.5L$.

The next step is to parameterise the shape of the vertical profiles of F_s. The parameters describing the shape of the profiles will then be linked through a multivariate regression to the boundary conditions such as current velocity, sailing speed, et cetera. Depending on the ambient conditions and overflow jet exit conditions, two distinct types of plumes can be distinguished: the near-bed density current and the seabed-detached plume. The shape of the vertical flux profile of both types of plumes is clearly different (Figure 9).

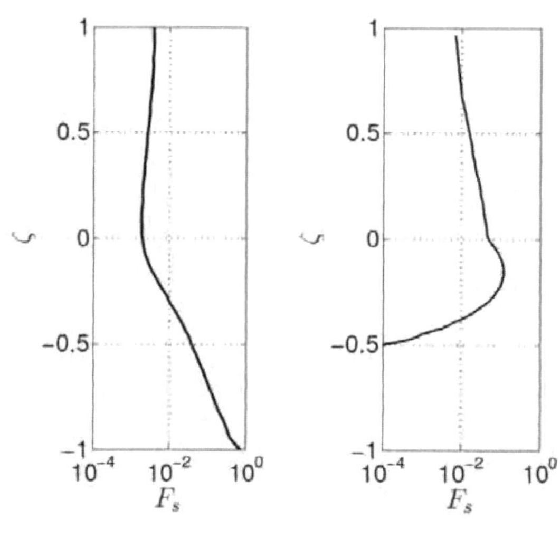

Figure 9: Vertical profile of F_s for two types of plumes: Near-bed density current type (left panel) and the seabed-detached plume (right panel).

In order to select which type of profile will occur, a preliminary estimate of the vertical position of the plume centerline at $x=x_p$ is required. For this purpose, the Lagrangian model for the trajectory of buoyant jets of (Lee and Cheung, 1990) and (Lee and Chu, 2003) is used as a starting point, with corrections based on regression analysis using the CFD results. In case the preliminary plume centreline is at $\zeta <$

-0.75, the plume is close to the sea bed and considered of density current type. If $\zeta > -0.75$, the plume is defined as type seabed-detached.

The shape of these profiles was then parameterised using either a Chebychev polynomial (density current) or a piecewise-linear (seabed-detached) approach.

The first type, the density current type, usually has a relatively smooth profile, and can be approximated using Chebychev polynomials, see e.g. (Lopez, 2001). In this method, a weighted sum of polynomials with order zero to n is considered (eq. 5). The coefficients ψ_i in the weighted sum are fitted to each case in the data set of CFD model-based plumes. Here, polynomials with n=3 provided sufficient capability of following the shape of the profiles:

$$F_s(\zeta) = \sum_{i=0}^{n} \psi_i \, T_i(\zeta) \tag{5}$$

where Ti are the Chebychev polynomials of the first kind and ψ_i are n+1 coefficients.

For the second type of plume, the seabed-detached plume, a step-wise parameterisation of the flux profile is proposed. The reason for the different parameterisation is the fact that this type of profile is often less smooth, with a sharp edge at the position of the bottom of the plume where the sediment concentration goes to zero rapidly. Fitting using Chebychev polynomials induces wiggles due to the sharp edge. In Figure 9, the step-wise parameterisation (during model training) for both types is shown.

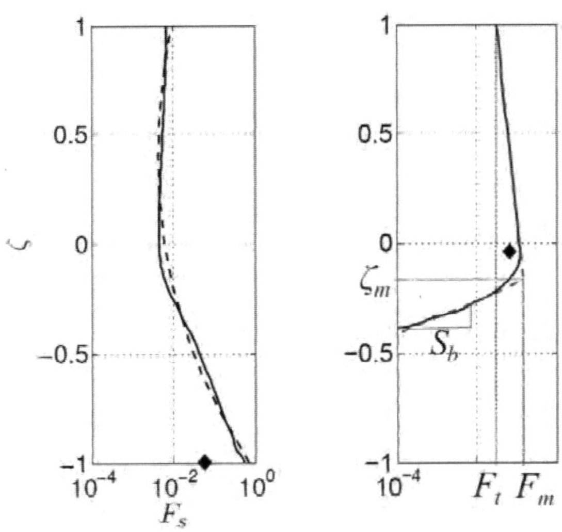

Figure 10: Vertical profile of F_s for two types of plumes. CFD results in full lines, parameterisations in dashed lines, preliminary plume centerline position in black diamond. Near-bed density current type with Chebychev parameterization (left panel). Right panel shows the 'detached' plume type with step-wise parameterisation (dashed line), defined by (co-) ordinates ζ_m, F_t and F_m and the slope S_b.

This gives a total of four parameters to fit to the data set, for both the Chebychev (ψ_i) and stepwise linear approach (ζ_m, S_b, F_t and F_m). These parameters are for that purpose defined as linear functions of the physical quantities of influence such as the solid discharge ($Q_{s,0}$), vessel draught (H_d), keel clearance (H_k), distance of overflow to stern (L_o), the ratio of outflow-to-crossflow velocity (λ) and the densimetric Froude number F_Δ:

$$F_\Delta = \frac{W_0}{g\, D \frac{\Delta\rho}{\rho_\infty}} \tag{6}$$

Where W_0 is the overflow exit velocity, D is the diameter and $\Delta\rho$ as defined in eq. (1).

Using multivariate linear regression, the relationship between the physical boundary conditions and the profile shape parameters was established, by fitting to a large set of CFD results. Here, this process is called model training.

3.3 Model Training

A training data set of 50 CFD simulations was used to relate the parameters to the different boundary conditions of the plume. These boundary conditions consist of dimensionless combinations F_{Δ}, λ, H_d/L_o, etc. The ranges of these conditions covered by the training data set determines the validity of the model, in this case: $1.2<F_{\Delta}<14.2$, $0.5<\lambda<4$, $0.07<H_d/L_o<0.26$ and $1<H_k/D<30.4$.

The Chebychev coefficients ψ_i (eq. 5) were found to depend mainly on F_∞ ($=F_{\Delta}$ / λ) and the ratio H_k/D. For each coefficient, a multivariate regression is fitted with these two dependent variables. The training data set cases are used for finding $\beta_{c,i}$ (3 x 4=12 coefficients):

$$\psi_i = \beta_{c,i,0} + \beta_{c,i,1} F_\infty + \beta_{c,i,2} \left(\frac{H_k}{D}\right)_m + \epsilon_{i,m} \qquad (7)$$

where i=0,...,3 is the number of the Chebychev coefficients, m=1,...,M, with M the number of CFD simulations in the data set, $\beta_{c,i,j}$ are the coefficients to fit and $\epsilon_{i,m}$ are error terms.

The parameters for the step-wise profile of the seabed-detached plumes were found to be best represented as a function of the following near-field plume conditions: F_∞, the ratio H_d/L_o and the ratio H_k/D. For each parameter, a multivariate regression is fitted with these three dependent variables. The training data set cases are used for finding β_d (4x4=16 coefficients):

$$(F_t, F_m, \zeta_m, S_m) = \beta_{d,0} + \beta_{d,1} F_{\infty,m} + \beta_{d,2} \left(\frac{H_d}{L_o}\right)_m + \beta_{d,3} \left(\frac{H_k}{D}\right)_m + \epsilon_m \qquad (8)$$

where, m=1,...,M, with M the number of CFD simulations (with seabed-detached plume) in the data set, $\beta_{d,j}$, j=1,2,3, are the coefficients to fit for each of the profile parameters (F_t, F_m, ζ_m, S_m). ϵ_m are error terms.

More details on the mathematical description and parameter settings of the parameter model can be found in Decrop (2015).

3.4 Validation

A second set of CFD simulations not used for the fitting of the coefficients of the multivariate regression was then applied as a validation data set. It turns out that the vertical profile of sediment flux in an overflow plume can be predicted reasonably well for most standard cases by this parameterised model.

Out of a total of 40 validation cases, 75% had a coefficient of determination (R^2) of 0.7 or higher. Two examples of comparison between CFD model results and parameter model results are shown in Figure 11. The parameter model is valid for a large range of boundary conditions and can after this validation exercise be used for generating source terms in a large-scale model, including the interaction between currents, vessel operation and plume behaviour.

3.5 Application in far-field plume models

The near-field sediment flux profiles generated by the parameter model can thus now be imposed as source terms in a large-scale plume dispersion model, in a coupled way. Effectively, an online coupling was established between the near-field parameter model and the TELEMAC code (EDF R&D, 2013) for solving large-scale hydrodynamics and sediment transport: the parameter model provides realistic sediment distributions for the sediment sources in the far-field model in TELEMAC, while the TELEMAC hydrodynamic solver provides water depth (i.e. ship keel clearance H_k), flow velocity and direction (changing in time) as an input for the near-field parameter model (Figure 12). The user provides the static variables to the parameter model at the start of the simulation: overflow discharge Q_0, diameter D and sediment concentration C_0, vessel length L_s and overflow position L_0. Vessel speed and course can be provided as an a priori-defined time series.

In this way, the sediment spill and its vertical distribution is fully adapted to the ambient conditions at any time step in the model.

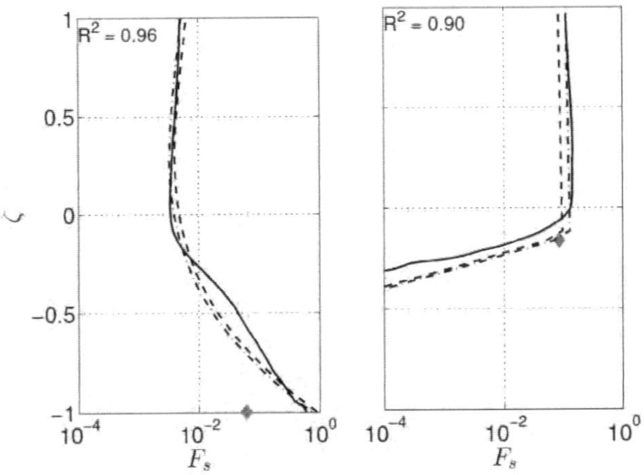

Figure 11. Two validation cases for the parameter model. Full lines represent vertical profiles of sediment flux from the CFD model, dashed lines are the results of the parameter model (before and after a corrector step ensuring the sediment flux continuity is respected).

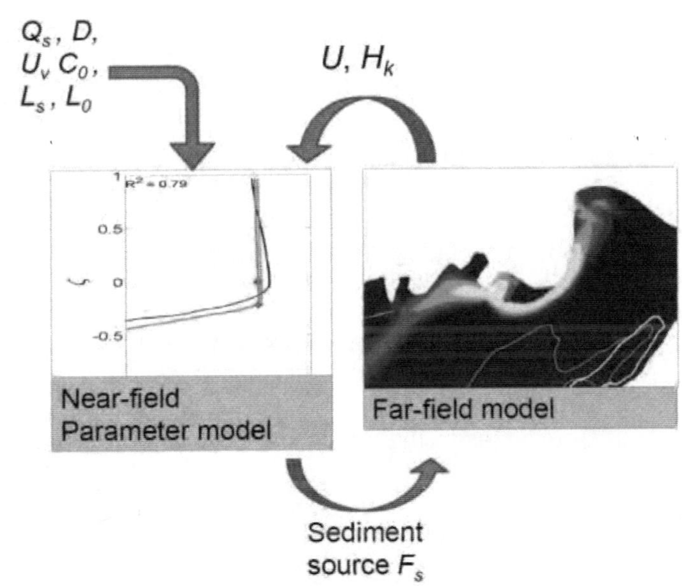

Figure 12. Schematic representation of the online coupling between near-field parameter model and far-field hydrodynamics/plume dispersion in TELEMAC.

3.5.1 Dredging scenarios

In a tender phase or during the phase of planning of the dredging works, predictions of the plume behaviour are calculated in a number of alternatives, to assess predicted compliance with environmental criteria. For example it can be decided which part of a dredging zone should be dredged during neap tide and which part during spring tide conditions.

4. DISCUSSION AND CONCLUSION

A highly-detailed CFD simulation model of near-field overflow has been developed. It was used to gain insight in the highly complex flows of water-sediment-air mixtures behind overflowing TSHD's. Further, it was used in the development of a faster parameter model of near-field overflow plume dispersion.

The sediment flux profiles generated by the newly developed parameter model can now be imposed as source terms in a large-scale plume dispersion model, where the parameter model inputs are coupled with the large-scale flow properties. In this way, the fraction of released sediments moving to the large passive plume is determined every time step. This is a significant improvement over the rather arbitrarily chosen constant value used in the past.

Currently, the grey-box model is applied by consulting engineers at IMDC during environmental impact assessment of port development and maintenance, both in scenario analysis and in a real-time plume forecasting system. Operational analysis of undesired sedimentation of the underwater work areas in between construction phases is another example of application.

In the future, the same approach - CFD simulations of the detailed processes and parameter model fitting based upon it - can be applied to other types of sediment spills from dredging and disposal activities related to port and navigation channel construction.

References

Bray R.N. (2008). "Environmental aspects of dredging. CRC Press".

de Wit L., van Rhee C. and Talmon A. (2014). "Influence of important near field processes on the source term of suspended sediments from a dredging plume caused by a trailing suction hopper dredger: the effect of dredging speed, propeller, overflow location and pulsing". *Environ. Fluid Mech.*, 1–26.

Decrop B. (2015). "Numerical and Experimental Modelling of Near-Field Overflow Dredging Plumes", PhD Thesis. Ghent University.

Decrop B., De Mulder T., Toorman E. and Sas M. (2015a). "Large-eddy simulations of turbidity plumes in crossflow". *Eur. J. Mech. BFluids*, 53, 68–84, doi: http://dx.doi.org/10.1016/j.euromechflu.2015.03.013.

Decrop B., De Mulder T., Toorman E. and Sas M. (2015b). "Numerical Simulation of Near-Field Dredging Plumes: Efficiency of an Environmental Valve". *J. Environ. Eng.*, 141(12).

Decrop B. and Sas M. (2014). "Challenges in the acoustic measurements of dredging plumes". Particles in Europe Conference, Esbjerg, Denmark 2014.

Decrop B., Sas M., De Mulder T. and Toorman E. (2014). "Large-Eddy Simulations of a Sediment-Laden Buoyant Jet Resulting from Dredgers Using Overflow". Int. Conf. Hydrosci. Eng.

EDF R&D (2013). "Telemac Modelling System, 3D hydrodynamics, Operating manual, Release 6.2".

Fischer H.B. (1979). "Mixing in inland and coastal waters". Academic Press.

Jirka G.H. (2006). "Integral model for turbulent buoyant jets in unbounded stratified flows Part 2: plane jet dynamics resulting from multiport diffuser jets". *Environ. Fluid Mech.*, 6(1), 43–100.

Lee J.H. and Cheung V. (1990). "Generalized Lagrangian model for buoyant jets in current". *J. Environ. Eng.*, 116(6), 1085–1106.

Lee J.H.W. and Chu V.H. (2003). "Turbulent jets and plumes: A Lagrangian approach". Springer.

Lopez R.J. (2001). "Advanced engineering mathematics". Addison-Wesley.

Manninen M., Taivassalo V. and Kallio S. (1996). "On the mixture model for multiphase flow". Technical Research Centre of Finland Finland. VTT Publications, 288.

Smith S.J. and Friedrichs C.T. (2011). "Size and settling velocities of cohesive flocs and suspended sediment aggregates in a trailing suction hopper dredge plume". *Cont. Shelf Res.*, 31(10), S50–S63.

Spearman J., De Heer A., Aarninkhof S. and Van Koningsveld M. (2011). "Validation of the TASS system for predicting the environmental effects of trailing suction hopper dredgers". *Terra Aqua*, (125).

NOMENCLATURE

Symbol [Unit]	**Definition**
c [kg/m³]	Sediment concentration
C_o , C_p [kg/m³]	Sediment concentration at resp. overflow and at dredging pump
D [m]	Overflow diameter
F_s [kg/m² s]	Sediment flux in the plume
F_t [kg/m² s]	Turbulent sediment flux
H_k [m]	Keel clearance
L_o [m]	Distance from overflow to stern
L_s [m]	Ship length
Q_o [m³/s]	Volume discharge at overflow
Q_p [m³/s]	Volume discharge at dredging pump
Q_s [m³/s]	Volume discharge in the plume
U, V, W [m/s]	Flow velocity components
U_v [m/s]	Vessel speed
U_0 [m/s]	Ambient flow velocity
W_0 [m/s]	Mean overflow exit velocity
x [m]	Horizontal distance from dredger
z [m]	Vertical distance from keel
φ_a [-]	Air volume concentration
$\varphi_{a,0}$ [-]	Air volume concentration at overflow exit
ρ_a , ρ_m , ρ_s , ρ_∞ [kg/m³]	Mass density of resp. air, plume mixture, sediment grains and ambient waters
ζ [-]	Normalised vertical coordinate

DIFFERENT ASPECTS OF DREDGING AND DISPOSAL WORKS, EXPERIENCES AND CHALLENGE IN PANAMA CANAL

Raúl Figueroa[1] and Melita Chin[2]

I. INTRODUCTION

In the maritime and logistic industry, the dredging activities are important for creating and safeguarding the navigation channel and ports in those areas. The Panama Canal, one of the most important maritime routes for the international commerce with a strategic geographic position, represents an economic, reliable and safety alternative. The Panama Canal Authority, in order to maintain the reliability of this maritime service, performs regular maintenance of the channel and invests in increasing its capacity and improving the services using state-of-art technology to strengthen the competitive position of the country and ensure its future viability (Figure 1):

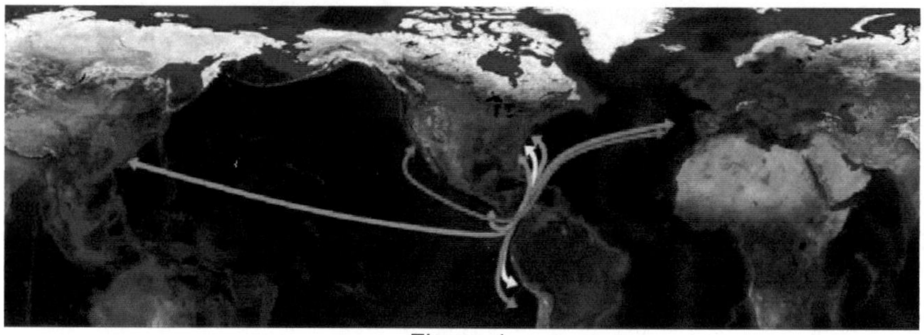

Figure 1

In this document, the intention is to describe the dredging works, related activities and aspects in the Panama Canal; that include the operational restrictions, maritime regulations, environmental and safety regulation, and others.

Remarkable excavation and dredging works were executed and completed during the last 17 years; being the most relevant capital dredging projects, those within the Expansion Program, that complemented the construction of the Third Set of Locks, opened for official operation on June 26, 2016. These projects included deepening and widening of the Navigation Channel required for the transit of the Neo-Panamax vessels.

In addition to this Expansion Program, the ACP continues with a comprehensive maintenance dredging program along the existing and the new reaches of the expanded channel, in the sea entrances and lakes, at the same time develops other projects to improve the efficiency of the Panama Canal operations and the quality of service, in response to projected increase in traffic, specially of Neopanamax vessels.

The main dredging challenge in the Panama Canal is to work during traffic operations, since the expeditious and safe vessel's transit is the priority. The work plan should consider this limitation, and its impacts, which could represent a reduction of production of the dredgers, increasing logistic and managing the related risks. Others challenges are the limited number of dredging equipment and its availability, the interface with other projects, navigation structures, maritime aid to navigation, geographical restrictions; the large range of geological formations; tides and critical lake levels; location and capacity of disposal sites and ACP regulation that could be relevant.

The unique experience and lessons learned from those projects feed the ACP´s knowledge and its ability to manage issues in the most efficient way, mitigating impacts and improving the dredging works, optimizing dredging methodology in the Panama Canal.

Panama Canal Authority
[1]RFigueroa@pancanal.com, Supervisor of Resource Unit of Dredging Division, Panama Canal Authority
[2]MChin@pancanal.com, Project Manager of the Project and Contract Administration Division, Panama Canal Authority

II. DREDGING AND DISPOSAL WORKS IN PANAMA CANAL

The review of the Panama Canal history evidences the improvement through the years according to the requirement of the international maritime commerce, in order to maintain this channel as a viable meeting point of world trade routes between East coast of U.S.A. and Asia, East coast of U.S.A and West South America, Europe and West South America and others. Some of the changes executed in the navigation channel, including the expansion program, were (Figure 2):

Deepening Projects	Reference level	Period	
Gaillard Cut and Gatun Lake	-12.19m MLL		1953
	-13.72m MLL		1967
	-14.63m MLL		1976
	-14.63m MLL	1979	1986
	-16.76m MLL	2002	2005
	-16.76m MLL	2006	2014
Pacific and Atlantic Entrance	-14.2 m MLW and MLWS	2004	2006
Atlantic Entrance	-16.1, 16.4 and 17.2m MLW	2010	2013
Pacific Entrance	-16.1, 16.3 m MLWS	2009	2013
Pacific Entrance (Widening)	-16.3 m MLWS	2015	2016
Atlantic Entrance	-16.4 m MLW		2017

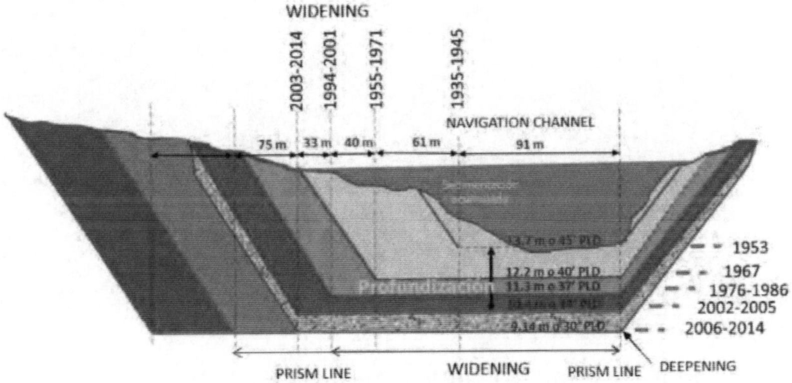

CROSS SECTION OF NAVIGATION CHANNEL

Figure 2

Considering the significant change of the navigation channel required for the expansion program, this document will briefly describe each dredging project executed by different Contractors and those performed with ACP´s equipment in this program as follow:

A. Dredging for the Expansion Program of the Third Set of Locks

The expansion program of the Panama Canal was a successful challenge, which provided since June 2016 to the maritime trade traffic a route with increased capacity using new locks for the NeoPanamax vessels. In order to provide the capacity, in terms of new locks, channels´depths and available widths for vessels bigger than Panamax, this expansion program required the timely execution and delivery of the following projects (Figure 3):

- Widening and deepening of the Atlantic Entrance and North Approach Channel to the Third Set of Locks.
- Widening and deepening of the Pacific Entrance and South Approach Channel to the Third Set of Locks
- Widening and deepening of Navigational Channels in Gatun Lake and Gaillard Cut.
- Excavation and dredging of the New Access Channel to Post-Panamax Locks - Pacific Side
- Increasing Gatun Lake's Maximum Operational Level
- Post-Panamax Locks – Pacific Side
- Post-Panamax Locks -Atlantic Side

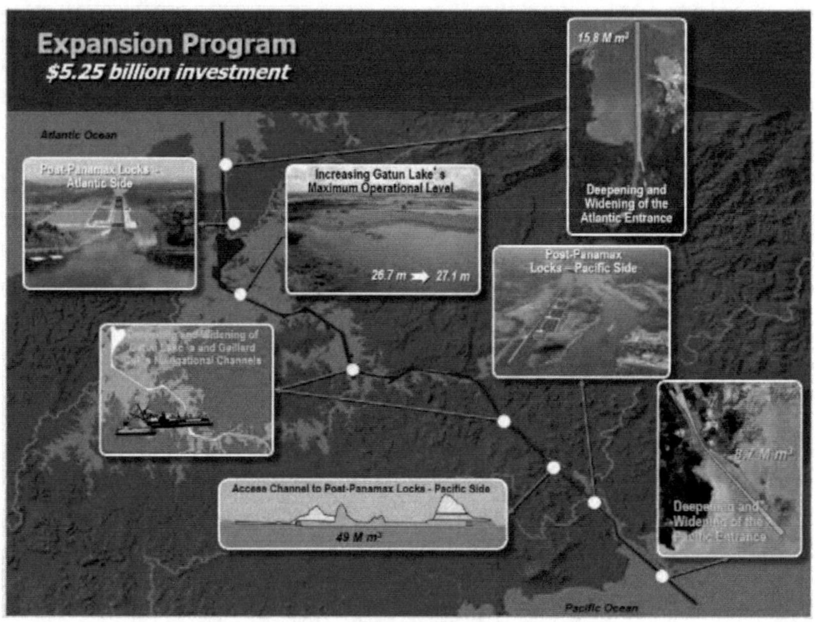

Figure 3

Projects under the Panama Expansion Program included a diverse set of activities and collateral works:

- Design, its revision, specifications and tenders.
- Site investigation.
- Topographic and hydrographic soundings.
- Socio-environmental activities as environmental impact studies, forest inventory, reforestation program, fauna rescue and relocation, archaeological survey, inspection and recovery of valuable items, stakeholder's management.
- Removal and relocation of structures, obstructions and debris.
- Construction of access and hauling roads.
- Preparation and construction of confined inland disposal sites.
- Erosion controls.
- Dry excavation, land based dredging and land-based drilling and blasting.

- Dredging and underwater drilling and blasting.
- Disposal of excavated and dredged materials.
- Movement and relocation of aids to navigation.

The general scope of the dredging projects, including ACP resources and contracts were:

- Widening and deepening of the Atlantic Entrance and North Approach Channel to the Third Set of Locks (figure 4):
  Dry excavation and dredged material: 16.8 Million cubic meters (Mm3)
  Extension: 13.8 km
  Equipment used for the project: 5 Trailing Suction Hopper Dredge (TSHD), 2 Cutter Suction Dredge (CSD) and 2 Backhoe dredgers (BHD), 5 split hull hopper barges, plough vessel.

Figure 4

- Widening and deepening of the Pacific Entrance and South Approach Channel to the third set of locks (figure 5):
  Dredged material: 8.7 Million cubic meters (Mm3)
  Extension: 14.3km
  Equipment used for the project: 2 TSHD, 2 CSD, 1 BHD, 1 Clamshell, 2 barges, 1 drill barge

Figure 5

- Widening and deepening of Navigational Channels in Gatun Lake and Gaillard Cut (figure 6):
  Dredged material: 25.8 Million cubic meters (Mm3)
  Extension: 49km
  Equipment used for the projects: 1 TSHD, 2 CSD, 3 BHD, 1 dipper handle dredge 2 drilling and blasting barges, 8 split hull hopper barges (chartered contracts and ACP Dredging Division)

Figure 6

- Access Channel to Post-Panamax Locks and plugs north and south (figure 7):
  Dredged material: 6 Million cubic meters (Mm3)
  Equipment used for the project: 2 TSHD, 2 CSD, 1 BHD, 1 dipper handle dredge, 2 drilling and blasting barge, 4 split hull hopper barges (Contractors and ACP)

Figure 7

After the completion of the expanded navigational channels, the area was increased in 38%.

The total cost of these projects reaches the $ 500 Million and where executed since 2008 till 2013.

B. Other Capital and Maintenance Dredging Projects to improve the capacity and support the Panama Canal Operations

Other projects executed to improve the capacity and allow growing operations of the Panama Canal are: deepening and widening of the navigation channels and turning basins, expansion of existing anchorage areas and create new ones; construction of docks, mooring stations and tie-up quay station.

Some of the following works, even with a reduced volume or extension, are cited due to its relevancy and faced challenging conditions.

These capital projects were developed in simultaneous with the maintenance of navigation channel, anchorage areas and docks which were executed framed by complex, balanced and dynamic schedule, which integrated efforts to be performed by ACP Dredging Division and Contractors. Some details of these projects are:

By Contractors:

- Complementary dredging works - Deepening of the Pacific Entrance and South Approach Channel to the third set of locks:

  Dredged material: 1.3 Million cubic meters (Mm3)
  Extension: 14.3 km
  Design depth: -16.3 m MLWS
  Equipment used for the project: 2 TSHD, 2 CSD, 1 BHD, 1 Clamshell, 2 split hull hopper, 1 drilling and blasting barge.

- Complementary dredging works - Deepening and Widening of the Atlantic Entrance:
  Dredged material: 6.6 Million cubic meters (Mm3)
  Extension: 13.8km
  Design widths: 300 m and 520 m
  Design depths: -16.4 m and -17.2 m MLW
  Equipment used for the project: 5 TSHD, 2 CSD, 2 BHD, 5 split hull hopper barges, plough vessel.

- Widening and maintenance of the Pacific Entrance and Balboa reach:
  Dredged material: 6.3 Million cubic meters (Mm3)
  Extension: 11.5km
  Minimum width: 300 m
  Equipment used for the project: 1 TSHD, 1 CSD, 2 split hull hopper barges.

- Maintenance and Deepening Dredging at Atlantic Entrance, Victoria Reach, Agua Clara Reach and Gatun Approach
  Dredged material: 2.1 Million cubic meters (Mm3)
  Extension: 7.2km
  Minimum Depth: -16.3 (Pacific) and -16.4 m (Atlantic)
  Equipment used for the project: 2 TSHD, 1 CSD, 2 split hull hopper barges, 1 BHD (ACP).

- Maintenance Dredging at Miraflores Approach and Balboa Reach, including Miraflores fueling facilities dock
  Dredged material: 1.8 Million cubic meters (Mm3)
  Extension: 4km
  Equipment used for the project: 1 TSHD

By ACP Dredging Division:

- Widening of Gaillard Cut
  Excavation and Dredged material: 29.5 Million cubic meters (Mm3)
  Extension: 12.5km
  Minimum width: 218 m
  Equipment used for the project: 1 CSD, 3 BHD, 1 dipper handle dredge, 2 drilling and blasting barges, 2 barges, land- based equipment and drilling and blasting.

- New Anchorages north of the Gatun Lake (Peña Blanca and Monte Lirio)
  Dredged material: 0.5 Million cubic meters (Mm3)
  Underwater clearing of submerged stumps Equipment used for the project: 1 CSD, 1 BHD, heavy drag barge

- Widening of Gamboa Reach and expansion of Gamboa Mooring Station
  Excavated and Dredged material: 11.9 Million cubic meters (Mm3)
  Extension: 7 Km
  Minimum width: 305 m
  Equipment used for the project: 1 CSD, 1 BHD, 1 dipper handle dredge, 2 drilling and blasting barges, 8 barges, land- based equipment and drilling and blasting.

- New Cocolí Mooring Station:
  Excavation and dredge material: 1.0 Million cubic meters (Mm3)
  Extension: 500 m
  Equipment to be used for the project: 1 BHD, 1 dipper handle dredge, 1 drilling and blasting barge, 4 barges, land- based equipment and drilling and blasting.
- Minor dredging works:
  Miraflores Fueling Facility dock basin.

Commercial (tourism) vessels dock basin.
Atlantic side Ferry's stations.
Atlantic Tug boat mooring station.

C. Operations and Maintenance of Navigation Channel

The Panama Canal has approximately 82 kilometer of navigation channel which correspond to 36 reaches (every straight section of the channel); are signaled with buoys, range towers and bank lights from North to South. These navigation aids attend the needs to safely transit through the Panama Canal without incidents or suspensions of regular operations, that means, 24 hours and 7 days a week. The availability and reliability of this channel for the transit of vessel requires monitoring and maintenance during all the year, dredging siltation and unclassified materials, including large scale rocks, within the navigation channel, mainly accumulated due to run-off, erosion, tides, landslides and hydrodynamic effects induced by transit operations (figure 8):

The maintenance work of the navigation channel was performed for years by human resources and dredging equipment of ACP. However, due to the conditions of the area, area restrictions, availability of equipment and/or optimization of the dredging work, simultaneous project development, the ACP shared this duties outsourcing some of them, by contracting dredging works or chartering dredging equipment. On the other hand, in very special cases, ACP has rented his equipment or provide support to contractors.

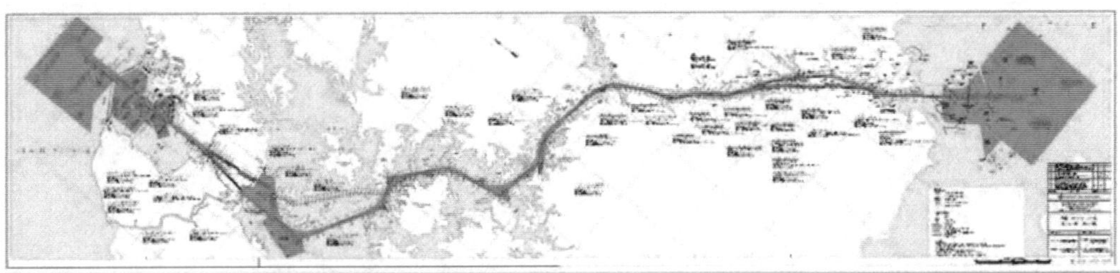

Figure 8

The navigation channel requires continuous maintenance to remove accumulated siltation alongside of the channel in order to provide sufficient clearance between the transiting vessel´s keel and the bottom, depending on the location, lake level (seasonal), type of vessels, siltation rates and condition of different areas, it is required to dredge with different frequencies those silted areas. Every year, the ACP requires dredge approximately 3.4 Mm3 of sediments, dredged by ACP Dredging Division and/or Contractors. This is conjunction with the attention needed in case of landslides (figure 9):

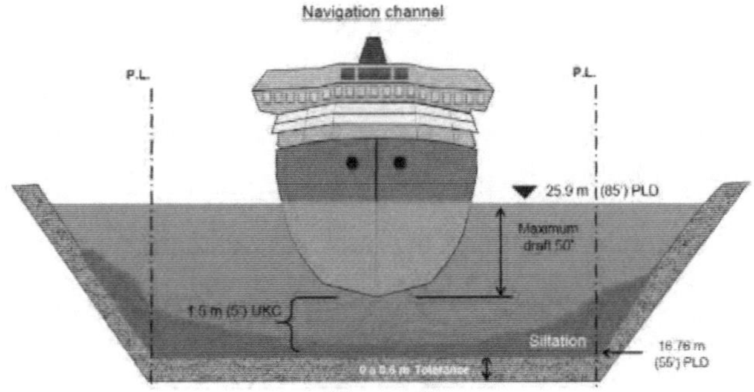

Figure 9

The dredging works activities require a very close monitoring and coordination 24/7 with Marine Traffic Control Center, who is in charge of coordinating and monitoring each vessel in transit, including launches, yachts or any other floating equipment, not only in the navigation channels, but those entering and leaving port basins, mooring statins or anchorages areas in the Panama Canal waters. In the canal, many launches and boats sail during the day and night that is one of the reasons to stay alert and keep in communication with the marine traffic office and pilots on board.

The dredging works inside and outside of the navigation channel need to be executed without interrupt the canal operation. This restriction includes all dredging activities executed by internal forces of ACP and by Contractors. Dredging operations are planned and coordinated with Port Captains and Traffic Management units in order to mitigate delays and to establish the estimated times and distances in regards to traffic.

The history of the Panama Canal operations registers only two event where the canal operations were totally suspended: the heavy rains on December 2010 named "La Purísima event" and the U.S.A. invasion on December 1989. None due to the dredging activities, for maintenance or deepening/widening of the navigation channel.

D. Regulations and Restriction in Panama Canal Waters

The Panama Canal is regulated by the ACP Operations Manual and the Pilot's Handbook based in the ACP Regulation on Navigation in Panama Canal Waters and International Regulations for Preventing Collisions at Sea, 1972 (72 COLREGS). Both manuals provide the rules and procedures that regulate the Canal operations.

The compliance of these regulations ensures the reliability, safety and efficiency of the Panama Canal operations. All of the vessel shall comply with these regulations, beginning with the compliance of the vessel certificates, inspections and any other requirements according to the characteristic of the vessel.

One of the principal rules is the delegated full authority for the vessel movements in Canal waters to the ACP Pilot to direct vessels 'maneuvers, coordinate activities and provide operation supervision over all personnel engaged in daily transit operations. Close and effective coordination among ACP transit management authorities and dredging operations units is key during planning and execution phases in order to achieve safe and efficient simultaneous operations.

The Operations Manual includes the procedures for dredging activities and navigation aids; establish the applicable restrictions depending on traffic conditions, the type of dredger, location of the dredging work that include the aspect like visibility, speed, window of work according to the transit schedule, signals and other conditions is evaluated by Transit Operation Division.

Other rules and procedures of the Operations Manual are aspects related with the transit, sail and dredging work in the Panama Canal Waters, as follow: communication, security and emergency, maritime safety, maritime accidents/incidents, pollution control, diving activities and others.

In complement to the Operations Manual and Pilot's Handbook, the safety, occupational health and socio-environmental norms and manuals shall be complied for the dredging works. Some other sensitive activities like towing of barges, plough operations, blasting and diving activities requires close monitoring and strict compliance of the rules and procedures.

III. DISPOSAL SITE AND MANAGE OF DREDGED MATERIAL

A. Land and underwater Disposal Site

The Panama Canal has several disposal site alongside of the navigation channel to be used for the improvement and maintenance of the channel in order to dispose the excavated and dredged materials, on inland and open-water disposal sites (figure 10). However, during the last years, most of these disposal sites were used for the expansion program of the Third Set of Locks and some of them should be reaching their maximum capacity after this program, hence, introducing an additional challenge for future marine projects.

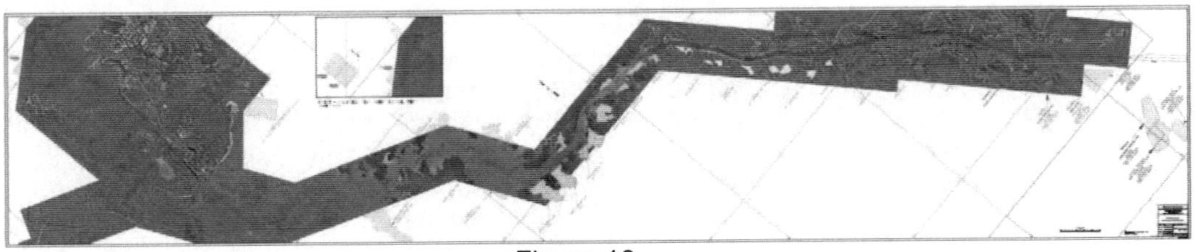

Figure 10

The ACP has classified the use of different disposal sites by the dredged material type to be disposed, the transportation distance and the available dredging equipment. This criterion was applied for the expansion program. It is an advantage for the future management of this material and the possibility of reutilization (figure 11):

Figure 11

The availability of disposal sites near to the dredge area is necessary to reduce de costs and optimize the production of the dredgers. Also, is a factor of the decision of the kind of dredge equipment to be used, due to the distance and the height difference, the dredge will require enough capacity and/or amount of pumps to reach the distance and high (figure 12):

Figure 12

Actually, the condition of available disposal sites for the maintenance and future improvement of the channel is a concern (figure 13). Some actions to solve this issue were performed, for example: increase the dike using the dredged material in the disposal, extend the area of the disposal, transfer the dredged material from the disposal to another area or for construction use; but it is still require a master plan for a solution, considering the annual siltation of 3.4 Mm3 to keep the navigation channel available and reliability for the transits.

Figure 13

B. Uses of Dredge Material

The Panama isthmus, close to be the younger piece of land of continent and the combination from seas to the mountains at the Continental Divide, has a variety of soils, ranging from fine sediments to huge boulders (figure 14):

Figure 14

After the expansion program, with approximately a total amount of 58 Mm3 material dredged, it places the ACP in a necessary condition to obtain an environmentally friendly solution, a medium and long term solution and at a reasonable cost.

Some of the solutions already applied to reuse of excavated and dredged materials are:
- for erosion control and landslides form channel´s embankments;
- for soil and slopes stabilization in the present location of North of PSA;
- sale of dredged material disposed in Velazquez, Mandinga South and Farfan disposal site;
- to improve and build new dikes on inland disposal sites (figure 15); also for underwater dikes to confine sediments within the disposal site;
- to improve and build roads;
- As enriched soil for grounds maintenance, farming and revegetation programs.

Figure 15

Even with these solutions, the ACP is still reviewing and analyzing others uses of these dredged materials, where could find a variety of soil and rock, because the Panama Canal has a particular geological condition in the navigation channel and the nearest area; different strata could found very close according to the boreholes done in the area (figure 16):

Figure 16

IV. EXPERIENCES AND CHALLENGE

A. Operational Restriction

One of the main restrictions for dredging works inside and outside of the Panama Canal is dredging with the transit of the vessel, avoiding any impact to the transit. Several restriction associated with the transit of the vessels is listed (figure 17):

- Keep the channel at least 122 m without any obstruction due to the dredging works, including the dredger, barges, cables, anchors, pipes, cardinal buoys, launches, tugboat, which may allow one-way traffic.
- Stop the hot works during the transit of the dangerous vessels (oil, etc.).
- The use of lights should not disturb the pilots during the transit of the vessels.
- Stop the diving works during the transit of the vessels if not keep the safety rules distance.
- Communication with the marine traffic office every time the dredger, launches or any other equipment enter the channel, or is relocated within it.
- Speed of the dredger or equipment avoiding impact to the vessels in transit and any other boat in the area.
- High restricted vessels may require the dredger to be relocated of the navigation channel.

Figure 17

B. Environmental and Safety Issues

The dredging works shall comply with the socio-environmental and safety international and national regulations.

Some of the safety issues presented during the dredging works are:

- Follow the safety procedures, for example: installation of floating pipelines.
- Boarding the dredger in a safety condition
- Use the safety equipment during all the working period
- Survey launch grounded
- Loss of the dredger engine
- Follow the procedures for works in confined spaces
- Comply with maritime regulations in terms of lights and day shapes to alert other vessels of the dredging conditions.

Some of the socio-environmental issues presented during the dredging works were:

- Restriction of overflow by dredgers, perform without permission or exceed the limits
- Not enough retention of sediment of the disposal site discharge
- Fuel spill without mitigation during the bunkering
- Not adequate wildlife rescue
- Stakeholders' management due to dredging and blasting operations.
- Water quality, air and noise standards.
- Recovery and conservation of archaeological and historical findings.
- Reforestation.

C. Lessons Learned

The world of the dredging works involve a lot of activities; depending on the location, the scope of work, weather, traffic and geographical restrictions, the available equipment, geology and client´s needs, that must be consider during planning and execution phases in order to reach the completion of the scope in time and within the cost, without accidents and incidents.

Beside the regular maintenances dredging the accumulated siltation in the navigation channel during several years from the administration of Panama Port Company (PPC) and some projects to improve the navigation channel, the most relevant challenge emerged with the expansion program. A lot of area spread along the waterway to be delivered and the corresponding volume to be excavated, and dredged, were required in a limited period, not only for the existing navigation channel, but for the new access for the new locks.

All of these projects allow contractors to learn beyond of our independent experiences, share them and live new ones during the development of the works. These experiences were complemented by those from consultants, who gave us guides to manage those mega projects. From this enriched environment, complex and valuable lessons learned arised, which we could summarize in the most important:

- The open and continuous communication within the involved party is key in dredging works that requires unstoppable 24 hours of work. Transfer the information between shifts is relevant for follow the work continuously and efficiently avoiding re-works or mistakes.

- The dredging works, especially with many dredgers in the same location, without interfering the regular transit of the vessel through the Panama Canal, in very restricted and traffic congestion areas, was a challenge. Some issues of communications and coordination impacted the production of the dredge at the beginning of the work. However, the close communication with Transit Operation Division (OPT), dredging operations teams and the time monitoring to remove the dredger to avoid the obstruction in the navigation channel solved the issue.

V. REFERENCES

1. Normas de Seguridad e Higiene Industrial. Autoridad del Canal de Panamá. 2011.
2. Manual de Seguridad y Salud Ocupacional. Autoridad del Canal de Panamá. 2007.
3. Manual de Operaciones. Autoridad del Canal de Panamá. 2018.
4. Pilot's Handbook. Autoridad del Canal de Panamá. 2002-2013.

HOW NAVIGABLE ARE FLUID MUD LAYERS?

by

Alex Kirichek[1], *C. Chassagne*[1], *H. Winterwerp*[1], *A. Noordijk*[2], *R. Rutgers*[2], *C. Schot*[2], *K. Nipius*[3] *and T. Vellinga*[1,4]

ABSTRACT

This paper gives an overview of the research that has been conducted to get a better understanding of the navigation in ports and waterways with fluid mud layers. In particular, the up-to-date review of reported full scale experiments that involve real vessels is provided. To study physical processes, the full scale experiments are accompanied by scaled experiments and numerical modelling. This combination provides a valuable insight into ship behavior with respect to different navigation conditions and physical properties of fluid mud. Another aspect of this paper are the surveying methods that can localize the fluid mud layers and potentially provide information about the strength of these layers. Some of these methods were tested on fluid mud produced by water injection dredging in the Port of Rotterdam. It was concluded that the new rheology-based method show a potential for understanding of strength development in fluid mud layers. Finally, some open research questions with respect to the applicability of the navigation through fluid mud are discussed.

1. INTRODUCTION

Safe navigation in ports and waterways should be insured by the port authorities and a good safety record is of utmost importance for the competitive position of the harbors. An important factor for safe navigation is the space that is available under a ship's keel, known as the under keel clearance (UKC). Sufficient UKC can be achieved by either setting a restriction on the maximum draft of incoming vessels or by maintaining the desired navigable depth with intensive dredging operations in silted areas. The first option is usually undesirable from an economic point of view and would result in a severe restriction on future development of larger and thus more energy-efficient ocean traversing vessels. The second option is favorable in general, although the cost and environmental impact of the required dredging operations to dispose the fluid mud can become quite substantial.

Fluid mud is typically deposited on the bottom of the shipping route when the net sedimentation rate is larger than the consolidation rate (Winterwerp, J. C. & van Kesteren, W. G. M. (2004)). Settling and consolidating in low-energy areas, fluid mud forms river and sea beds. Fluid mud consists of a highly concentrated suspensions of sediment particles combined with microbial slimes. These slimes can be seen as a network of polyelectrolytes that keep the sediment particles in suspension, and fluid mud can be seen as a visco-elastic fluid. Navigation in ports and waterways with fluid mud seabed can be challenging due to several factors. The fluid mud layer cannot be reliably detected by traditional acoustic methods. The interpretation of the measured acoustic data in ambiguous since the position of the seabed on the acoustic charts is not clear. Another challenge for navigation in muddy areas is the generation of internal waves (undulations). The controllability and maneuverability of a vessel can be hindered by such waves in case of a ship navigating in a close vicinity of a water-mud interface. In particular, the amplitude of the undulations affects the rudder and propeller efficiency as was observed during the both in-situ and laboratory experiments.

[1] Department of Hydraulic Engineering, Faculty of Civil Engineering and Geosciences, Delft University of Technology, the Netherlands, o.kirichek@tudelft.nl
[2] Department of Asset Management, Port of Rotterdam, the Netherlands
[3] Rijkswaterstaat PPO, the Netherlands
[4] Department of Environmental Management, Port of Rotterdam, the Netherlands

In 1970th, it was recognized that sonar measurements are not reliable in areas with substantial mud layers and, therefore, are no longer normative. On the basis of full scale experiments in the Port of Rotterdam, in Bangkok and along the coast of Suriname, it was found that densities up to 1200 kg/m^3 had a tolerable influence on maneuverability. The "Nautical Depth" was defined accordingly. The illustration of this concept is shown in Figure 1.

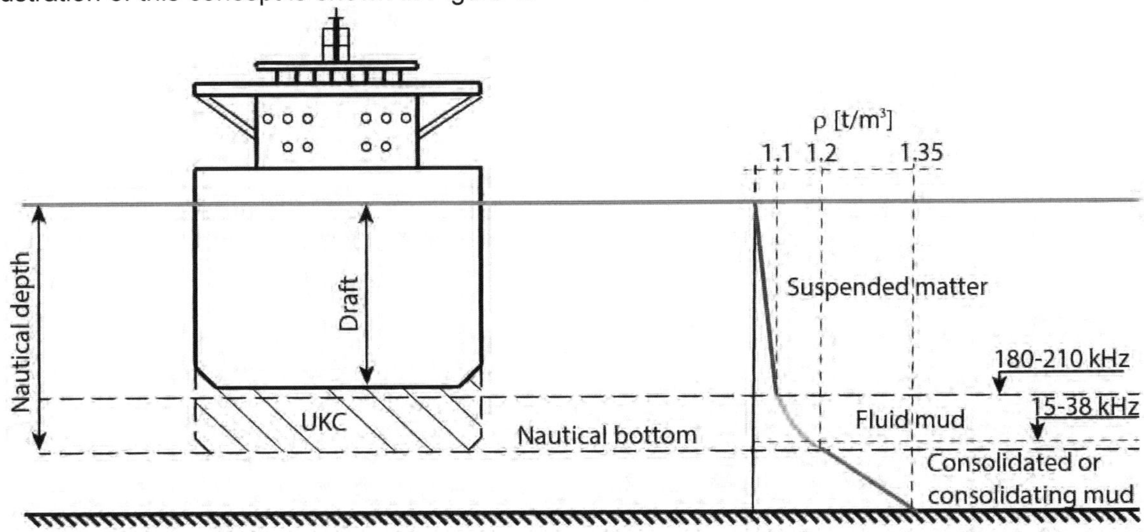

Figure 1: The nautical bottom concept, which was developed at the density limit of 1.2 t/m^3. Adapted from from Nederlof, L. (1978).

In 1980th, a series of full-scale experiments was conducted at the Port of Zeebrugge. The water-mud interface, that could be detected by the high-frequency (180-210 kHz) echo-sounder, was chosen as the reference for the UKC. The experiments were conducted with positive and negative UKC, that correspond to the level above or below the depth that was detected by high-frequency echo-sounder. It was concluded that fluid mud with a low density may lead to a modified sailing behavior, but does not give rise to dangerous situations. Together with outcomes of towing tank studies, the results of the full-scale experiments led to a change of the nautical bottom density criterion from 1150 to 1200 kg/m^3., see Table 1.

Port	country	criterion	value	dimension
Rotterdam	the Netherlands	density	1200	kg/m^3
Zeebrugge	Belgium	density	1200	kg/m^3
Bordeaux	France	density	1200	kg/m^3
Nantes-Saint Nazaire	France	density	1200	kg/m^3
Dunkirk	France	density	1200	kg/m^3
Avonmouth	the UK	density	1200	kg/m^3
Yangtze	China	density	1250	kg/m^3
Liang Yungang	China	density	1250-1300	kg/m^3
Yianjing Xingang	China	density	1200-1300	kg/m^3
Bangkok	Thailand	density	1200	kg/m^3
Paramaribo	Suriname	density	1230	kg/m^3
Emden	Germany	yield stress	100	Pa

Table 1: Criteria for nautical bottom from McAnally, W. H. et al. (2007)

Currently, the following definition for nautical bottom is used for navigation in muddy areas: "The nautical bottom is the level where physical characteristics of the bottom reach a critical limit beyond which contact with a ship's keel causes either damage or unacceptable effects on controllability and maneuverability" (PIANC (2014). Unlike other ports, the Port of Emden uses the yield stress as the criteria for the nautical bottom since 2005 (Wurpts, R. (2005)). In Emden, the adopted criteria above which fluid mud is navigable is a yield point of 100 Pa. It is thought that this parameter is related to the specific dredging method or the slime properties is this port. It remains to be investigated how the dredging properties affect the composition of the mud, and hence its rheological properties

The overviews of full scale and scaled experiments and modelling for ships navigation in the channels with fluid mud layers can be found in Vantorre, M. (1994), Delefortrie, G (2008) and Delefortrie, G. & Vantorre, M. (2015). The present article provides an additional discussion on surveying methods that can be used for monitoring the fluid mud layers.

2. FULL SCALE EXPERIMENTS

The first reported full scale experiments were conducted in the Port of Rotterdam in 1975 (Nederlof, L. (1978) and Nederlof, L. (1979)). The 318t deadweight oil tanker SS Lepton (see Figure 2) was sailing within the Europort area at Hoek van Holland in and out the port. During the tests maneuverability data and the density depth profiles were recorded. The average mud layer thickness in the investigated area was about 1.15 m. The maneuvers were done at UKC of 6 % of the draft with respect to the depth indicated by the 210 kHz echo-sounder. Although the pilots experienced the entry as 'difficult', a comparison of the maneuvering behavior of the Lepton with that of other incoming tankers shows that in fact it was not more difficult than other large ships and even easier than other vessels. This conclusion was also supported by the measurement of the speed of the tanker, maximal percentage of available rudder and propulsion which were analyzed during the experiment.

Figure 2: The 318t deadweight oil tanker SS Lepton was a subject of investigation for the full scale experiment at the Port of Rotterdam in 1974. Here, SS Lepton is filled with water enters the Port of Rotterdam. Observers on board carefully follow the sailing behavior of the ship. The Lepton was 350 m long and 55 m wide, the draft was 20.90 m, the average throat clearance was 1.60 m, the average mud layer was about 1.15 m. The photo is taken from Nederlof, L. (1979).

Other well-reported full scale experiments were conducted in Zeebrugge in 1986-1988. Seventeen full scale tests of three types were carried out with the trailing suction hopper dredger Vlaanderen XVIII (van Craenenbroeck, K. & Vantorre, M. (1991)). Short engine maneuvers (acceleration/deceleration tests), constant power maneuvers, yawing tests at zero speed by means of bow thrusters were conducted in the outer harbor of Zeebrugge. The maneuvers were performed with keel clearances from -0.35 to +3.0 m with respect to the depth indicated by the 210 kHz echo-sounder. The rotation tests were conducted at a keel clearance from -0.3 to -0.4 m. During the trials, the presence of internal waves on water-mud interface was confirmed. The solitary stern internal waves were detected by the 210 kHz echo-sounder. An internal wave of about 2 m height was detected by surveyor vessels during the passage of a deep-drafted OBO carrier.

Two full scale trials were conducted in the Port of Delfzijl in 2013-2015. The first trial was conducted with the general cargo vessel CSL Rhine at positive UKC of 14% and larger with respect to mud-water interface that was detected by 210 kHz echo-sounder. The results showed that the maneuvering and propulsion behavior were influenced by the mud layer at UKC smaller than 18%. The second trial was carried out with the hopper dredger Geopotes 15. The tested UKC was between -5% and +17% draft to the depth detected by 210 kHz echo-sounder. During the test it was concluded that during the navigation with the UKC of +14% with respect to the mud-water interface caused less disturbance than to CSL Rhine at the same UKC. It was concluded that the maneuverability was much more favorable with fluid mud layers of 2.5 to 3.5 m thickness in combination with a negative UKC than with a positive UKC from +1% to +13%. At the UKC from +10% to 20% the ship-induced internal waves hindered the propeller and the rudder of the vessel at low viscosity mud layers. It was reported that the vessel's maneuvers were in line with simulation studies (Verwilligen, J. et al (2014) and Barth, R. et al. (2015)).

3. SCALED EXPERIMENTS

Scaled modeled tests were performed to get better understanding of ship's maneuvering in muddy environment. These test were coupled to mathematical maneuvering models to quantify the effect of the mud layers on the controllability of the ship. The scaled experiments with real mud can be problematic because of the time effects (e.g. settling, consolidation) of mud. Therefore, a dense viscous fluid is typically used to mimic the fluid mud in a two layer system. The physical properties of an artificial mud (density, viscosity) are chosen to be close to the ones of the fluid mud.

One of the first scaled tests, that were carried out to investigate the effect of fluid mud on maneuvers, were conducted with a 1:82.5 scale model of a tanker that was equipped with rudder and propeller, sailing above or in contact with a dense fluid layer (Sellmeijer, R. & van Oortmerssen, G. (1983)). A mixture of chlorinated paraffin and kerosene was used to simulate the fluid mud layer. Two density values, 1140 kg/m^3 and 1240 kg/m^3, were chosen to mimic the real mud conditions at the Port of Rotterdam during the winter and summer periods, respectively. The tested dense fluid layer thickness, from 1.35 m to 3.85 m, was varied. Both positive and negative UKC with respect to the water-dense fluid interface were tested by changing the water level in the basin. Two types of experiments were carried out: free running tests were conducted to evaluate the total effect of the mud on the maneuvers of the vessel and the captive experiments were carried out to predict standard maneuvers by means of mathematical models, which describe the ship motion in a horizontal plane. The effects of mud on squat and trim of the tanker were analyzed. It was observed that the tanker becomes slower with the UKC of 3-5 % of the draft above the dense fluid. However, further reduction to negative values of UKC makes the tanker less slow in its maneuvers. Furthermore, the presence of dense fluid on the bottom tends to reduce the steady motion and to accelerate the dynamic motions. One important observation was the internal waves that occurred in the water-dense fluid interface when a ship is passing. The amplitude of these internal waves (see Figure 3) increases with the thickness of the dense fluid layer and with decreasing fluid density and affect the propeller efficiency as was observed during the free running trials.

A series of model scale tests were carried out in a wave flume with a tanker model of scales 1:100, 1:70 and 1:55 (Brossard, C. et al. (1991)). The model was equipped with sensors for measuring squat, trim and tractive force. The goal of the experiments was investigating the resistance and squat variations above an artificially composed mud layer. The tested mud layer had density gradients over the depth. Three type of mud was tested: with low, intermediate and high gradients. The layers with different yield stresses were used in the study. The undulations on the water-mud interface had been observed in this study.

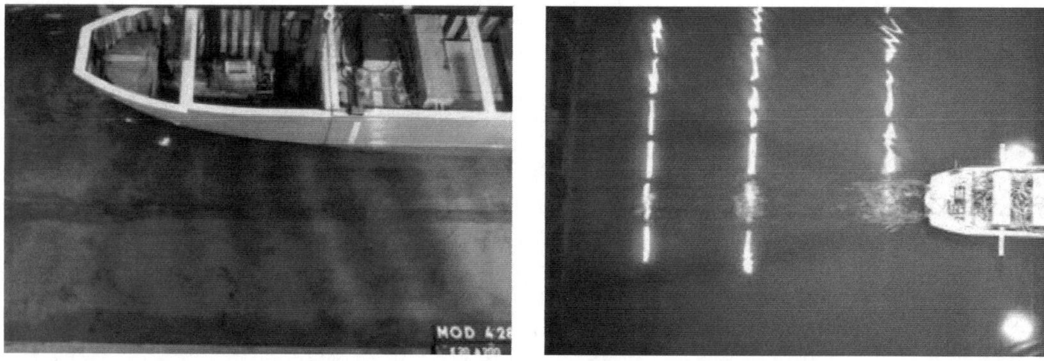

Figure 3: Propagation pattern of internal waves of the water-dense fluid interface for viscosity 0.002 Pa s (left) and 0.03 Pa s (right). The photos are taken from Delefortrie, G. & Vantorre, M. (2015).

The self-propelled tests were conducted with scale models of an LNG-tanker and a hopper dredger along a guiding rail above water-mud system with a dense fluid mud layer. Unstable rudder behavior and poor propulsive efficiency was observed if due to a combination of initial UKC, squat effects and internal waves, the ship's keel is in contact with both fluids (Vantorre, M. & Coen, I. (1988)). A series of captive test was conducted for different simulated maneuvers (Delefortrie, G. et al. (2005)). During the tests, longitudinal and lateral force components, vertical motion, rudder parameters and propeller parameters were measured. A mixture of chlorinated paraffin and petrol was used to mimic the mud. Mathematical maneuvering models were developed for fluid type of densities 1100-1250 kg/m^3 and of viscosities 0.03-0.46 Pa s, dense fluid layer thickness of 0.75, 1.5 and 3 m, for 1:80 and 1:75 scaled models sailing with UKC of -12 to +21 % relative to water- dense fluid interface and the speeds between 2 knots astern and 10 knots ahead.

4. MONITORING METHODS

Traditionally, the reflections of acoustic signals are used to determine the positioning of water-bed interface. The emitted acoustic pulse propagates through the water column and reflects back from the bottom of the waterways. The distance from the acoustic source to the reflecting surface is proportional to the travel time of acoustic waves in the water column. In muddy navigational areas, different frequencies of the emitted signal are employed. Standard low frequencies (15-38 kHz) and high frequencies (180-210 kHz) signals are used to provide information about water-fluid mud interface (lutocline) and bed-fluid mud interfaces, respectively. The former typically exhibits a strong contrast. The latter is often inconsistent due to a weak density gradient within a fluid mud layer that plays an important role in the reflection of emitted acoustic signals (Kirby, R. et al. (1980). Therefore, other measuring techniques have been proposed for the monitoring of the water-bed interface in muddy navigational areas. These techniques are typically based on the physics of the scattered and transmitted gamma-radiation, acoustic and optical backscatter, or through mechanical devices. One of the most accurate methods so far is based on scattered and transmitted gamma-radiation. This nuclear method is typically used to determine the density of the water column, including fluid mud layers. The instruments that are based on acoustic and optical backscatter typically measure the concentration of suspended particles in the water column. All the non-acoustical methods have several common drawbacks. Due to the nature of the profilers, the spatial resolution of these tools is limited to 1D vertical profiles (see Figure 4). Thus, the interpolation between the measurements to obtain a spatial map is generally done by combining these methods with acoustic sounding. Another important disadvantage of these methods is their intrusive nature of surveying. The measuring tools have to be in a direct contact with fluid mud layers in order to provide quantitative characterization.

Even though it has been recognized that the surveying methods that are based on rheological parameters are the most suited for nautical purposes, the in-situ measurement of these parameters is a challenging task. Two strategies can be found to determine these parameters. The first one is that samples are taken in-situ and analyzed in the laboratory. The second one is to use in-situ instrumentation. Both strategies are discussed below.

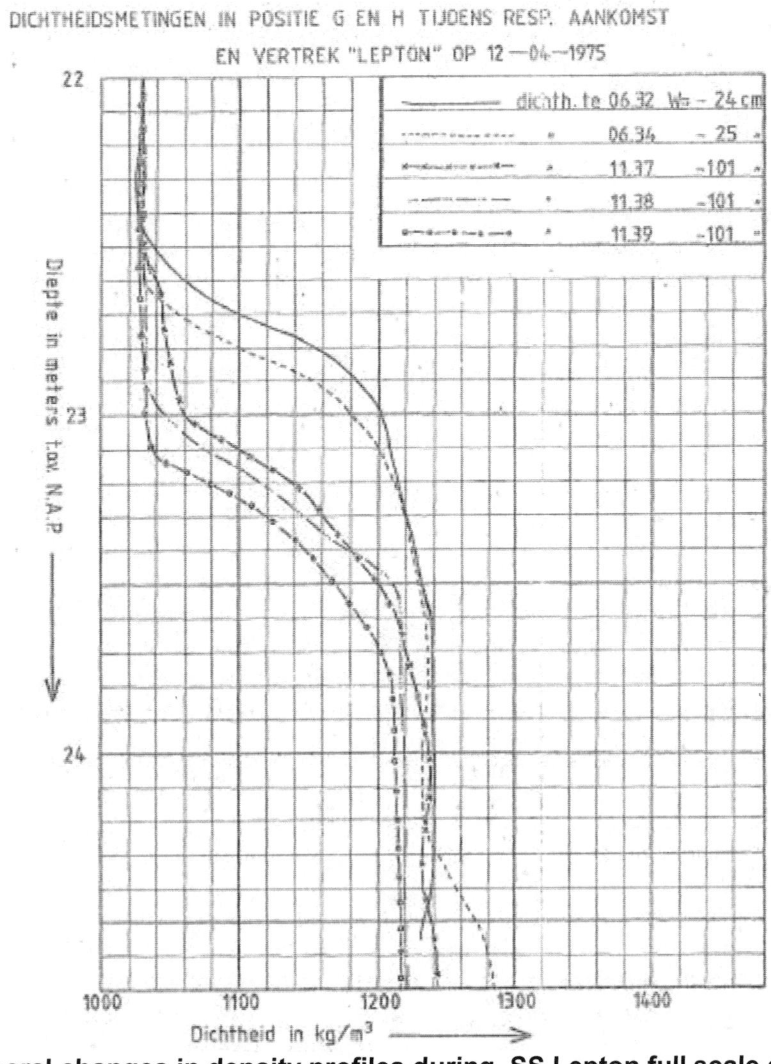

Figure 4: The temporal changes in density profiles during SS Lepton full scale experiment in 1975. The diagram is taken from Rijkswaterstaat (1977).

4.1 Laboratory

Rheological properties can be routinely determined in the laboratory, for instance by the vane-type tests or rotational rheometers. These laboratory methods measure the resistance of fluid mud samples to flow in response to applied shear forces. This can be done by controlling the shear rate, $\dot{\gamma}$, or shear stress, τ, that gives the flow curves (see Figure 5) for different mud samples. Two mechanical behaviors of mud can be deduced from a traditional flow curve: elastic and viscous. The elastic behavior is conventionally observed at very low shear strains rates. As soon as the strength of the soil matrix weakens, mud starts to flow exhibiting viscous behavior. In this mechanical state, deformations correspond to the rate of deformation. The shear stress, at which this soil matrix start to deform (or to flow) is conventionally called the yield stress. Typically, the yield stress, τ_y, and dynamic viscosity, μ_∞ , can be obtained for different mud samples from measured flow curves. Two additional parameters are required for reconstructing the complete flow curve: the Bingham stress, τ_B, and the initial deferential viscosity, μ_0. As can be seen on Figure 5, a given density does not necessarily correspond to a unique yield stress. This implies that the relation between density and rheological parameters should be carefully studied in order to assess the best set of parameters required to characterize the nautical bottom.

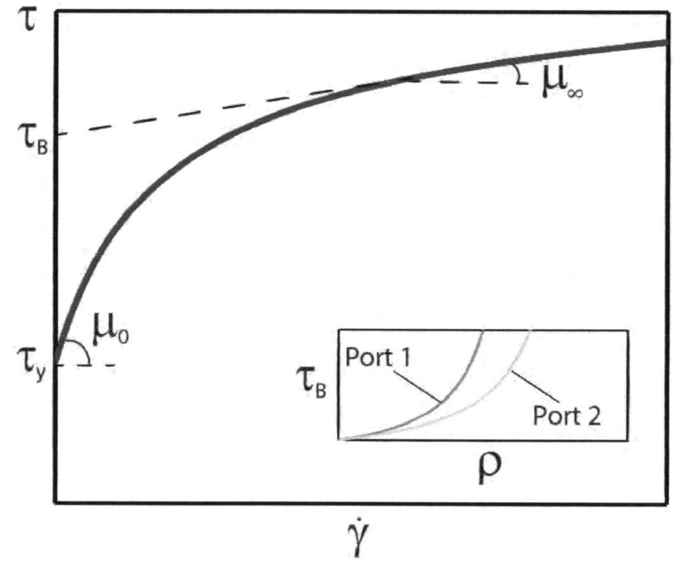

Figure 5: The sketch of the flow curve with rheological parameters.

Due to the complexity of fluid mud, the shear stresses exhibit a non-linear relationship with the density, ρ. This can be explained by the thixotropic behavior (deformation, history and time dependence) that is illustrated in Figure 6. Therefore, sampling and measuring procedures, followed by data processing and interpretation of the measurements, have to be standardized by means of recognized practical protocols (Claeys, S. et al. (2005)).

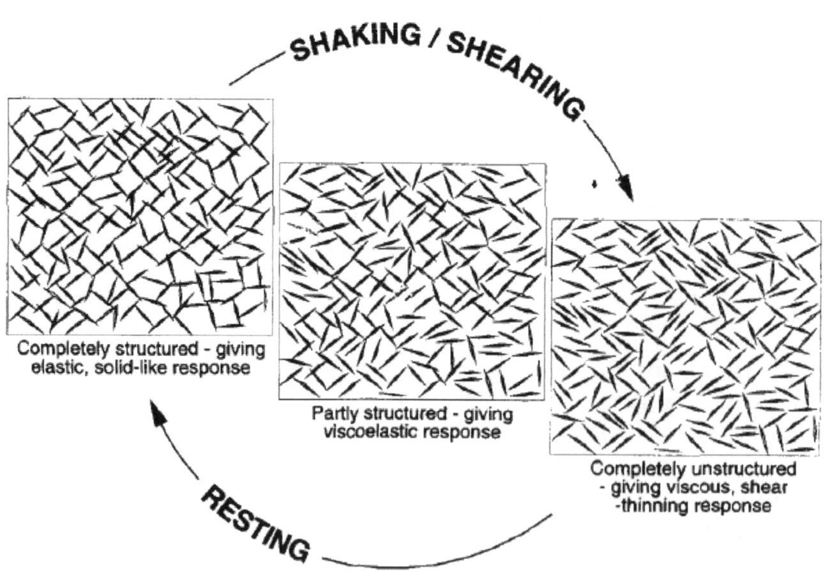

Figure 6: Breakdown of the thixotropic structure. Taken from Barnes, H. A. (1997).

The three next sections concern in-situ instrumentation.

4.2 Free falling cone penetrometers

The physics of a free falling cone penetrometer is based on recording the penetrometer's acceleration/deceleration for getting the information about the cone penetration resistance. Using a calibration procedure, the rheological properties of the site-specific mud can be related to this cone-end resistance. In this way, the water-mud column profiles can be mapped by using this type of instrument.

Figure 7: Examples of rheology-based methods. From left to right: lab rheometer (Anton Paar), free falling cone (Graviprobe), tunning fork (Rheotune), towing object (Rheocable)

One of the advantages of this type of equipment is that the vertical positioning of the penetrometer can be derived from the recordings of accelerometer. In general, an accelerometer is more accurate in indicating the depth than standard pressure sensors.

4.3 Tuning fork profilers

The tuning fork profilers are based on the recording of the amplitudes that are triggered by mechanical vibrations at different frequencies. These recordings can be used to get the information about the yield stress and viscosity of mud. For this purpose, an accurate calibration procedure, that requires laboratory rheological measurements, is necessary. The tuning fork profilers can provide sufficiently accurate rheological properties in the areas of low sediment concentrations.

Figure 8: Location of the experiment in the 8th Petroleumhaven. The man-made pit has dimensions of 500 m × 200 m.

4.4 Towing objects

The depth level of the towed object is defined by the viscosity discontinuity between consolidated and fluid mud. The high viscous forces in the consolidated mud and heavy cable weights attached to the towed object assure the cable to position itself on the interface between fluid and consolidated mud unless a critical towing velocity is exceeded and the towed object starts floating in the water above the fluid mud layer. In the case of the Rheocable, the continuous measurements of the electrical resistivity value is used to verify whether the cable is on the seabed or floating above it. The depth level of the towed object is defined by a pressure sensor on the seabed measuring the hydrostatic pressure (Druyts, M. & Brabers, P. 2012).

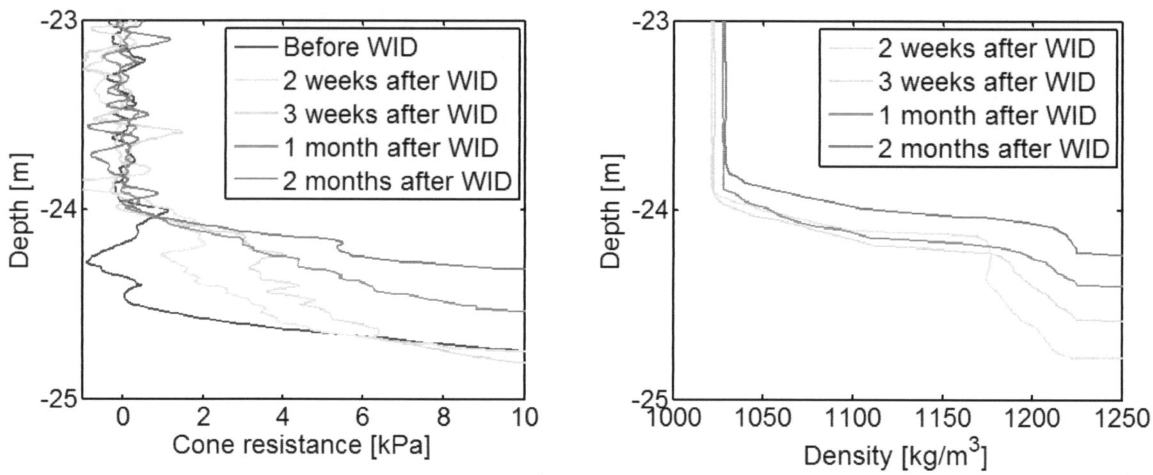

Figure 9: Fluid mud development in time, that is measured by Graviprobe (left) and DensX (right).

4.5 Comparison of density- and rheology-based monitoring

In order to compare the density and rheology based monitoring, the water injection dredging method was employed in the 8th Petroleumhaven at the Port of Rotterdam (see Figure 8). The water injection dredging (WID) was used for liquefying the top layers of the sediment around a man-made pit in the river bed so that the mud would flow into the pit (500 m $\times$ 200 m). The pit collected the fluid mud layer up to 1.5 m thickness from the surrounded area. The Graviprobe and DensX was used to capture the development of the fluid mud layer in the pit over a period of 2 months. The results of the monitoring are presented in Figure 9.

The recordings of the Graviprobe are shown as the cone resistance measurements as function of depth. The dark blue profile represents the measurements in the pit conducted before mobilizing of fluid mud into the pit by water injection. The cyan, green, magenta and red profiles show the fluid mud strength development in the deepening after 2 weeks, 3 weeks, 1 month and 2 month after the water injection, respectively. The same colors are used to show the density profiles that were measured with DensX. From the measurements, it can be concluded that the density of fluid mud develops faster than its strength.

Figure 10 shows the comparison of the Rheocable and Graviprobe surveys which were conducted after the WID in the 8th Petroleumhaven. The results provide a good agreement regarding navigable depth that is given by the methods. The high frequency echo-sounding measurements (200kHz) are included for comparison.

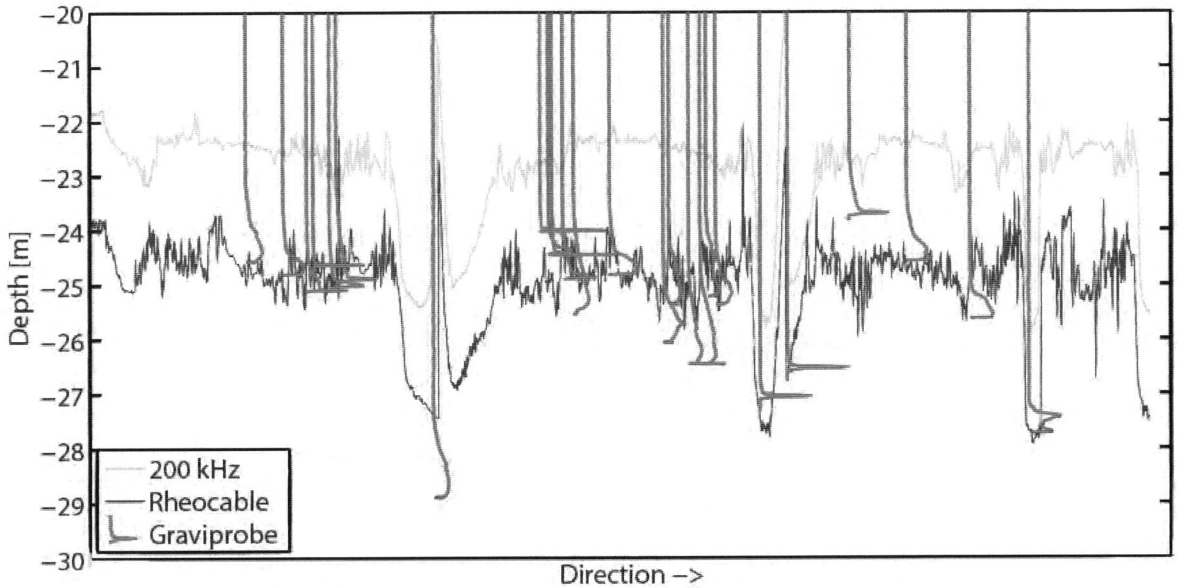

Figure 10: Correlation of the high frequency echo-sounding (200 kHz), Rheocable and Graviprobe measurements in the 8th Petroleumhaven.

5. The way forward

The safe navigation requires a new reliable and universally accepted criteria for the definition of the nautical bottom. Until now, this criteria was related to the density of fluid mud layers. Some steps have already been taken toward the new definition, for example in the Port of Emden that considers the yield stress as the parameter for the nautical bottom.

Already several decades ago it was recognized the importance of improving the parameters used in the definition of the nautical bottom. At that time, in-situ rheological experiments were practically impossible. This forced the community to adopt an alternative parametrization, based on density. Over the years, there has been a whole new set of equipment put on the market meant to study the rheological properties of the fluid mud layer (e.g. Graviprobe, Rheotune, Rheocable). These instruments are currently tested in different ports, and it remains to be investigated if the parameters derived from the experimental results are compatible with one another. More advanced 3D acoustic methods are needed for mapping the fluid mud layers. Open questions are:

1) How to relate acoustic measurements to the rheological properties of fluid mud?
2) Can rheological point-measurements be used to calibrate acoustic mapping?

Measuring rheological parameters, and in particular yield stresses, require a well-thought and universally accepted protocol, as these parameters are strongly history-dependent and lead to thixotropic effects.

3) What would be an accepted protocol for measuring rheological parameters?

The measured in-situ rheological parameters have, for calibration purposes, to be compared to samples analyzed in the laboratory. Care should be taken during the sampling, storage and analysis of these samples. The composition of the mud samples (in terms of mineralogy, biological and organic matter content) should be related to the rheological properties and the thixotropic behavior of mud.

4) How can mud composition be incorporated in the rheological parameters?

The mud properties (density, viscosity, yield stress, mud layer thickness) are time dependent because of consolidation processes in the fluid mud layer. Moreover, mud layers can be mobile. More knowledge is needed to understand the time dependence of the fluid mud layers.

5) What is the development of the mud layers over time and what is the link between density and rheological parameters?

The rheological parameters are to be incorporated in CFD models, to predict the forces acting on the ships, in accordance with the suggestions made by Delefortrie, G. & Vantorre, M. (2015).

6) How to successfully implement the complex rheological behavior of fluid mud in CFD models?

After answering the questions given above, it is necessary to engage all the key stakeholders into the discussions on a global acceptance of new criteria for the nautical bottom.

7) How to obtain and promote an international implementation of the new criteria related to the nautical bottom definition?

All these points are currently under investigation in different research groups world-wide.

References

Barnes, H. A. (1997). Thixotropy – a review. Journal of Non-Newtonian Fluid Mechanics, 70.

Barth, R., van der Made, C. J. A. W., Bourgonjen, L., van Dijken, J., Vantorre, M. & Verwilligen, J. (2015). Manoeuvring with negative underkeel clearance: 2nd full scale field test in the Port of Delfzijl. Proceedings of the 4th International Conference on Ship Manoeuvring in Shallow and Confined Water with Special Focus on Ship Bottom Interaction.

Brossard, C., Delouis, A., Galichon, P.. Granboulan, J. and Monadier, P. (1990). Navigability in channels subject to siltation. 22nd International Coastal Engineering Conference.

Claeys, S., De Schutter, J., Vantorre, M. & Van Hoestenberghe, T. (2011). Rheology as a survey tool: "We are not there yet". Hydro International 15(3).

Claeys S., Staeiens, P., Vanlede, J. Heredia, M., Van Hoestenberghe, T., Van Oyen, T. & Toorman, E. (2005). A rheological lab measurement protocol for cohesive sediment. Book of abstracts: 13th International Conference on Cohesive Sediment Transport Processes.

Delefortrie, G., Vantorre, M. & Eloot, K. (2015). Modelling navigation in muddy areas through captive model tests. Journal of Marine Science and Technology 10.

Delefortrie, G. & Vantorre, M. (2015). Ship manoeuvring behaviour in muddy navigational areas: state of the art. Proceedings of the 4th International Conference on Ship Manoeuvring in Shallow and Confined Water with Special Focus on Ship Bottom Interaction.

Delefortrie, G (2008). Manoeuvrability of deep drafted vessels in muddy navigation areas. PIANC magazine 133.

Druyts, M. & Brabers, P. (2012) Nautical depth sounding – the Rheocable survey method. International Hydrographic Review.

Kirby, R., Parker, W.R., van Oostrum, W.H.A. (1980). Definition of the seabed in navigational routes through mud areas. International Hydrographic Review, Monaco, LVII (1).

McAnally, W. H., Teeter, A., Schoellhamer, D., Friedrichs, C., Hamilton, D., Hayter, E., Shrestha, P., Rodriguez, H., Sheremet, A., & Kirby, R. (2007). Management of Fluid Mud in Estuaries, Bays, and Lakes, Part 2: Measurement, Modeling, and Management. J. Hydraul. Eng, 133(1).

Nederlof, L. (1978). "Varen boven slib" in het Rotterdamse haven- en rivieren-gebied. Gementewerken Rotterdam. 98.21-R7332.

Nederlof, L. (1979). Varen door een waterlaag met meer zwevende slib deeltjes. S. en W - 46ste jaargang, 22, 461464.

PIANC (2014). Harbour Approach Channels - Design Guidelines, Report 121, PIANC, Brussels.

Rijkswaterstaat (1977). Varen boven slib. 98-20.W77.31.

Sellmeijer, R. & Van Oortmerssen, G. (1983). The effect of mud on tanker manoeuvres. The Royal Institution of Naval Architects, Spring Meetings, paper no. 7.

van Craenenbroek, K. & Vantorre, M. (1992)_Navigation in muddy areas: establishing the navigable depth in the Port of Zeebrugge. Terra et Aqua, 47.

Vantorre, M. & Coen, I. (1988). On sinkage and trim of vessels navigating above a mud layer. The Royal Society of Flemish Engineers, Harbour Congress.

Vantorre, M. (1994). Ships behavior and control in muddy areas: the state of the art. Third international conference on Maneuvering and Control of marine crafts.

Vantorre, M. (2001). Nautical bottom approach: application to the access to the harbour of Zeebrugge. Hansa 138(6).

Verwilligen, J., Vantorre, M., Delefortrie, G., Kamphuis, J., Meinsma, R. & van der Made, K.-J. (2014). Manueverability in proximity of nautical bottom in the Harbour of Delfzijl. PIANC World congress.

Winterwerp, J. C. & van Kesteren, W. G. M. (2004). Introduction to the Physics of Cohesive Sediments in the Marine Environment: Developments in Sedimentology, vol. 56, Elsevier.

Wurpts, R. (2005). 15 Years Experience with Fluid Mud: Definition of the Nautical Bottom with Rheological Parameters. Terra et Aqua, 99.

FEHMARNBELT – A NEW GREEN LINK BETWEEN GERMANY AND DENMARK

by

Anders Bjørnshave[1], Ian Sehested Hansen[2], Victor Magar[3] and Juan C. Savioli[4]

The world longest immersed tunnel is to be built in northern Europe and will connect Scandinavia with Germany, thus closing a major gap in the European transport network, reducing the risk of shipping collisions, energy consumption and creating a new region in Europe, while also fostering the development of new nature and recreational landscapes by Working with Nature concepts.

The proposed landscaping is considered a "win-win" situation where the surplus of 19 million m^3 of dredged material will be used to create new landscaping features that will add a positive impact to the project in the form of providing new engineered "natural" landscape elements. This will bring new natural, environmental and recreational values to the area; rehabilitating it to an extent after being blighted by engineering projects in the past.

1. PRESENTATION OF THE PROJECT

The Fehmarnbelt Fixed Link, the proposed eastern route between Germany and Denmark, has been planned for a long time to create a rapid connection between the two countries with a faster and shorter link, see Figure 1. It will also close a major gap in the Scandinavian-Mediterranean corridor, part of the European transport network. Once opened, the journey for freight trains between Hamburg and Scandinavia will reduce by 160 km.

The Fehmarnbelt Fixed Link is a joint Danish and German transport infrastructure project across the Fehmarnbelt. Denmark is responsible for the planning, construction and operation of the Fehmarn Fixed Link. To carry out this task, the government of Denmark has established the company Femern A/S, which is 100% own by the Danish State, represented by the Danish Ministry of Transport.

The initial feasibility studies of the project were already conducted in the mid-90s and eventually followed by the Danish-German treaty signed in 2008. The project was given the go-ahead in Denmark in 2015 and is expected to be approved in Germany by mid-2018. The project includes an immersed tunnel that will run 18 km under the Fehmarnbelt and connect Puttgarden (Fehmarn island, Germany) to Rødby (Lolland island, Denmark). The maximum water depth will be about 30m and the tunnel has an estimated total construction cost of EUR 7 billion.

The design of the link and its land reclamation was developed with the support of both environmental consultants e.g. DHI and COWI, the technical consultants Ramboll-Arup-TEC and the landscape architects Schønherr.

[1] Femern A/S, Head of Department, Environment. abj@femern.dk, Denmark,
[2] DHI, Head of Projects, Ecology Department. ish@dhigroup.com, Denmark,
[3] Ramboll Environ, Principal. vmagar@ramboll.com, United States
[4] DHI, Head of Projects. jcs@dhigroup.com, Malaysia

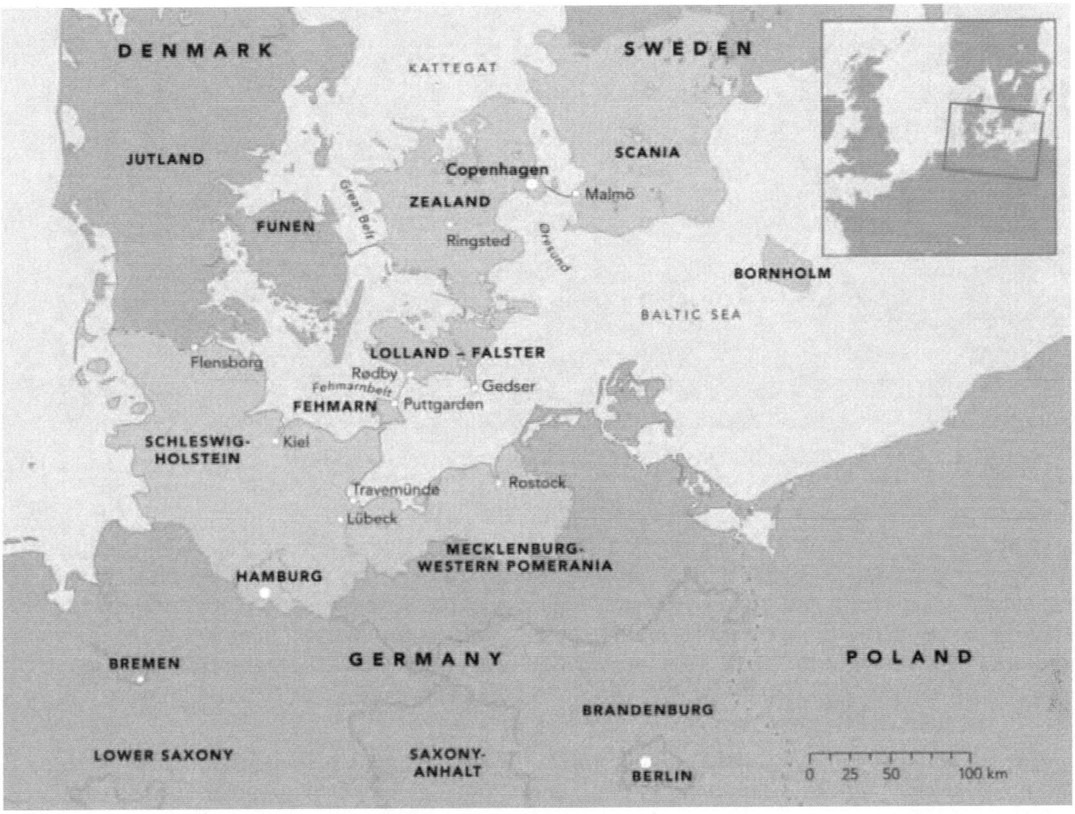

Figure 1. The upcoming Fehmarnbelt Fixed Link will be the eastern alternative to the western route from Hamburg over Funen and Zealand. The new green link will save time as well as energy.

The Fixed Link is a combined road and railway connection carrying two lanes of road traffic and a single, high speed rail track in each direction. When complete, it will be the third largest marine infrastructure project in southern Scandinavia. It will supplement the first one, the Great Belt Link between the Danish islands of Funen and Zealand, which opened in 1998, and the second, the Øresund Link, which opened in 2000 and connects Denmark's capital, Copenhagen with Sweden's third largest city Malmö. Ten million people will then be brought closer together, enlarging each country's growth region into one major regional centre. This will bring numerous opportunities for development, exchange of culture and business, trade and education etc.

2. THE JOURNEY TO THE IMMERSED TUNNEL

The immersed tunnel will be constructed by placing tunnel elements in a trench dredged in the seabed, see Figure 2. The proposed methodology for trench dredging comprises mechanical dredging using Backhoe Dredgers (BHD) up to 25m and Grab Dredgers (GD) in deeper waters. A Trailing Suction Hopper Dredger (TSHD) will be used to rip the clay before dredging with GD. The excavated material will be loaded into barges and transported for beneficial use to the inshore reclamation areas where it will be unloaded by small BHDs. Some 19 million m^3 of sediment will be handled.

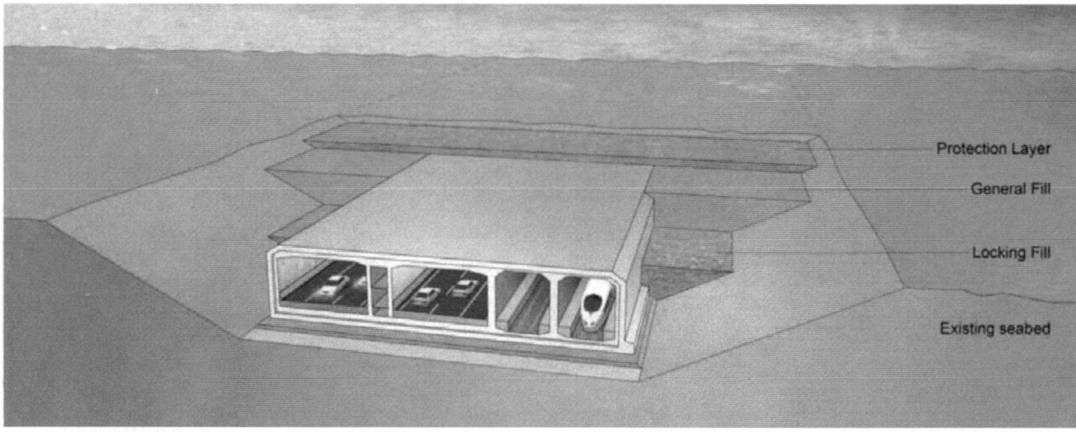

Figure 2. Cross section of dredged trench with tunnel element and backfilling

A bedding layer of gravel will form the foundation of the elements. Each element will initially be kept in place by depositing locking fill followed by general fill. Finally there will be a stone layer on top to protect against damage from grounded ships or dragging anchors. The protection layer and the top of the structure will stay below the existing seabed level, apart from near the shore. However, at the very nearshore area, the seabed will be raised at the coastal locations to incorporate the protection layer over a distance of approx. 500-700m from the proposed coastline. Here the protection layer will be an extended armour rock layer.

Reclamation areas are planned to run along both the German and Danish coastlines to use the dredged material from the excavation of the tunnel trench. The size of the reclamation area on the German coastline has been minimised. Two larger reclaimed areas are planned on the Danish coastline. Before the reclamation takes place, containment dikes are to be constructed some 500m out from the coastline.

The cut & cover sections of the immersed tunnel passes through the shoreline reclamation areas on both the Danish and German sides.

3. KEY REQUIREMENTS

One key requirement of this project is that it has to be designed and constructed in harmony with the landscape and its nature areas, thereby offering the opportunity to:

- Re-establish/re-generate some of the environmental values lost during the construction of major dikes and reclamation works in the early 1900's;
- Use the opportunity to create new landscapes;
- Incorporate the link into the landscape without visual harm;
- Make sustainable options a requirement.

Another requirement was that the Fixed Link poses minimal navigation risks to the important international navigation route connecting the North Sea and Baltic Sea. Finally, that the Fixed Link should have a minimal impact on the environment, particularly the water- and salinity exchange through the project area.

4. UNDERSTANDING THE ENVIRONMENT

To ensure the fulfilment of the above mentioned requirements it is necessary to understand the existing environment, not only to minimise the potential impacts of the project but also to identify win-win opportunities to meet the project requirements.

A detailed two-year survey programme was undertaken using a combination of fixed measurement stations, vessel and air surveys, as well as modelling works to study the important hydraulic and ecosystem components. These included water quality, benthic flora and fauna, fish, marine mammals and seabirds. With this information, it would be possible in the first instance to identify sensitive areas. The understanding of the environment also included analysis of the potential impacts on the sensitive areas.

The effects on higher trophic levels such as birds and marine mammals were assessed based on the outcome of these simulations. Ecological modelling was used to quantify the impacts arising from spilled sediment on water quality and benthic flora. The ecological model describes the relationship between dispersed sediment spill concentrations, light availability and primary producers, between nutrients, as well as the interrelationship and inter-specific competition between three distinct groups of producers: pelagic phytoplankton, benthic macroalgae and rooted vegetation.

The simulation of a realistic construction scenario for the entire construction period and of the permanent operation period of the link demonstrated that the Fixed Link could be built with only minor temporary and permanent impacts. A thorough understanding of the ecological aspects and coastal processes was important in order to identify options to re-generate the environment as well as to define sustainable coastal protection options. Understanding of the environment was a key element for the subsequent stages of the project and for selection of the preferred solution: the immersed tunnel that would provide the option of the beneficial use of dredged material to re-generate the landscape.

Some characteristic elements of the Fehmarnbelt area are described below:

4.1 Geology

The landscape of the area was shaped by the ice masses mainly during the last Ice Age. Since the final retreat of glaciers from the south-western Baltic area, the Fehmarnbelt has been characterised by highly variable sedimentary processes and environments, when the outflow from the Baltic Sea to Kattegat through the Great Belt and Øresund changed position several times. The present day topography and bathymetry was formed by the last Ice Age, which ended about 10,000 Before Present [B.P.], with varying water levels in the period.

The upper subsoils in the Fehmarnbelt consist mainly of glacial meltwater sand covered by clays and topped by post glacial marine sand, gyttja and peat. Beneath these layers are mostly glacial tills (also called boulder clay or moraine clay) of different types with local pockets of meltwater sand and silt, see Figure. 3. Deeper layers are chalk and paleogene clay that are older than the Quaternary period.

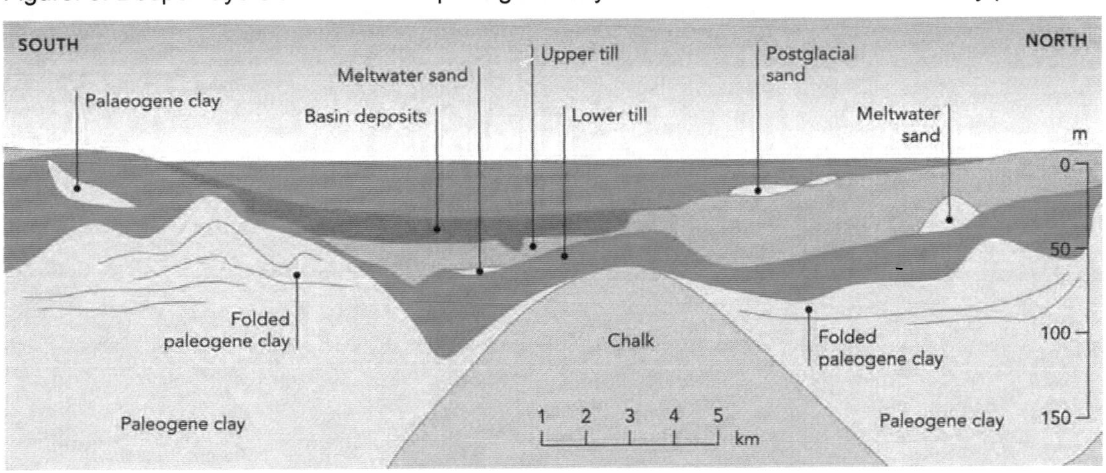

Figure 3. The geology of the Fehmarnbelt

4.2 Hydrography

The Fehmarnbelt is part of the transition area between the central Baltic Sea and the North Sea. The flow and stratification in the Fehmarnbelt is highly related to water exchange between the North Sea and the central Baltic Sea. The upper water strata in the Fehmarnbelt consist of brackish water from the central Baltic Sea, which, close to the surface, flows through the Belt Sea and continues up into Kattegat. A layer of water with higher salinity from the North Sea forms a lower layer.

4.3 Environment and Nature

The Fehmarnbelt is a very dynamic area in terms of water exchange and sediment transport, which forms different types of seabed substrate and forms, as well as coastal features such as cliffs and beaches.

In the shallow areas, the benthic flora is dominated by different flora communities determined by the water depth (light penetration) and substrates (mud, sand, hard bottom). Red algae communities (Fursellaria, Phycodrys, Delesseria species) are replaced in less shallow water by brown algae (Fucus Sp). On water depths over 20m, algae communities are rare. In wave-protected lagoons and bays, red algae are replaced by eelgrass. Blue mussels dominate along the Danish coast and are succeeded by amphipods (Bathyoireia and mussel communities (Corbula, Artcica species) in deeper water.

In terms of fish, the Fehmarnbelt is an important route for migrating cod, herring and silver eel, as well as a spawning area for a number of fish species, including cod and flatfish in general. In the Fehmarnbelt area, three species of marine mammals occur regularly: the harbour porpoise; the harbour seal; and the grey seal. The harbour porpoise use the Fehmarnbelt as a transport corridor whereas the seals do not, although seal haul-out sites are located over 20 km away both west and east of the alignment.

Birds are a special issue as the Fehmarnbelt is an important migration route both for north-south and east-west migration. Many species of water birds use the areas during either winter or summer. For that reason a number of protected habitats and bird areas have been designated. These are the so-called Natura 2000 areas according to EU legislation, see Figure 4.

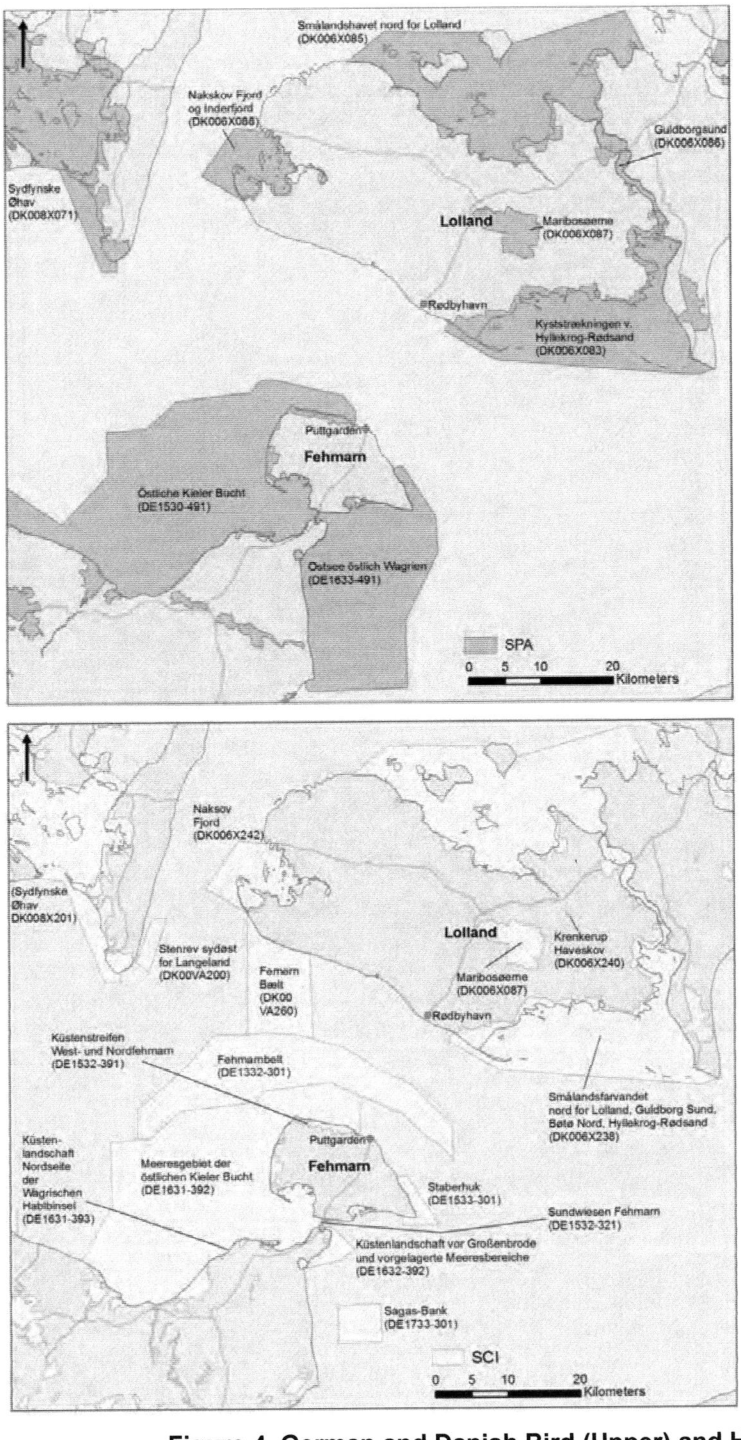

Figure 4. German and Danish Bird (Upper) and Habitat (Lower) areas

4.4 Human activity

Archaeological investigations conducted by the client reveal that the proposed alignment has been a human migration corridor over the last 6,000 years. DNA tests of archaeological finds of goat remains show that ancient human migration from central Europe to Scandinavia used the Fehmarnbelt as the main transport corridor. Today the Fehmarnbelt is one of the heaviest trafficked waters in the world because it is the main entrance to the Baltic Sea. It can be seen from the map (Figure 5) that ferries between Puttgarden and Rødbyhavn contributed with 35,000 crossings in 2013.

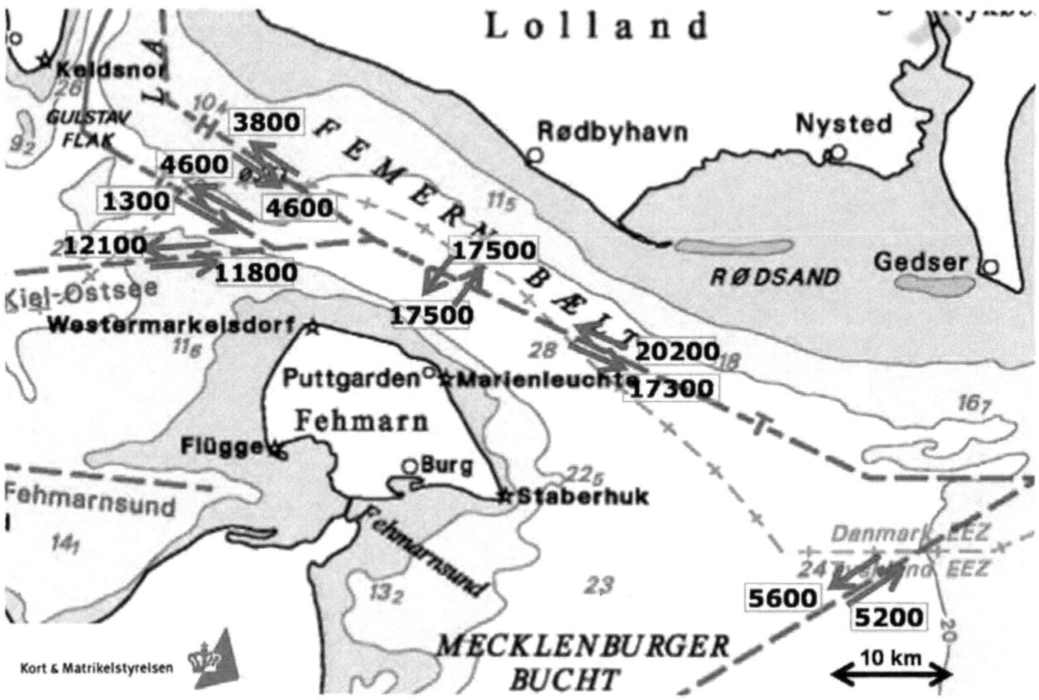

Figure 5. Number of ship movements in 2013 on main sailing routes in the Fehmarnbelt area determined on the basis of AIS data, ref. /20/

5. PROJECT DESIGN TO BENEFIT NAVIGATION AND NATURE

The project was designed by a multidisciplinary team consisting of engineers, architects, biologists, environmental engineers, etc. The design was based on the project objectives defined in early stages of the project and the understanding of the existing environment. These, together with extensive stakeholder engagement, worked towards a design that meets the stakeholder objectives and identifies win-win opportunities. Three options were initially proposed for the link. These were;

- a cable-stayed bridge,
- a bored tunnel, and
- an immersed tunnel.

The immersed tunnel was identified as the preferred option in the light of engineering, environmental, navigational and economic considerations. A tunnel provides safe navigation conditions since it avoids potential damage and associated oil spill resulting from ship collisions with piers and other obstacles.

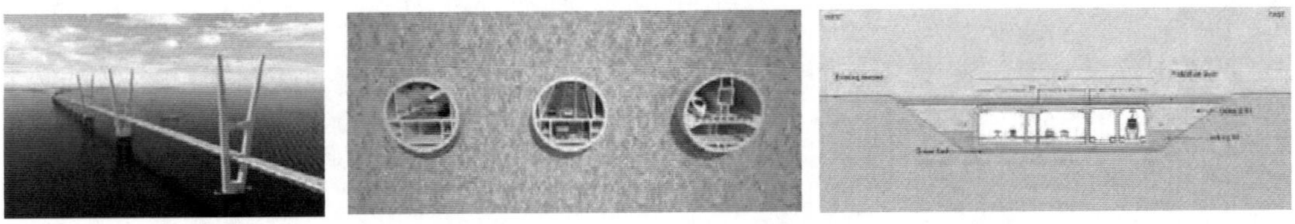

Figure 6. Left; Cable-stayed bridge with two main spans of 724 m and navigation clearance above the sea of at least 66 m. Middle; Cross section, the bored tunnel consists of three tubes. The railway tube has an internal diameter of 15.2 m, while the two road tubes' diameter are 14.2 m. Right; The immersed tunnel consists of 79 standard elements (approx. 9x42x217 m) and up to 10 special elements (approx. 13x45x39 m).

With respect to the landscape impacts of a tunnel and bridge solution differ. As opposed to the tunnel, which is submerged, a cable-stayed bridge has a clear visual impact on the entire area. On the environmental scale, the bridge would entail permanent barrier effects e.g. on the hydrographical conditions of the Baltic Sea and on bird migration in the area. The impact assessment on the surrounding Natura 2000 areas proved that the tunnel produces significantly fewer environmental conflicts than the bridge.

One reason for deselecting the bored tunnel solution was the uncertain time horizon for the possible re-usage of the bored material for land reclamation purposes, due to its slow dewatering process. Other reasons were that the bored tunnel had a larger (environmental) footprint on Fehmarn and significantly higher greenhouse gas emissions.

The immersed tunnel provided a "win-win" solution because the 19 million m^3 would provide an opportunity for the beneficial use of dredged sea-bed material from the tunnel trench were seen as a great and feasible opportunity to create new landscapes and re-establish some of the historical features that were lost due to coastal protection and flood mitigation works carried out in the past.

5.1 New reclamation areas add nature and recreational values

The new landscape will be shaped as a streamlined area along the existing coast to an extent similar to that of the existing Rødbyhavn harbour. This will ensure that there will be no additional blocking of the flow through the strait.

The new landscape, see Figure 7 will extend approx. 3 km west of the harbour and 3.5 km east of the harbour. The extent of the area is decided mainly by the volume of the surplus sediment that can be absorbed by the landscape. The western part of the reclamation area is designed with the purpose of serving recreational values whereas the eastern part, although it accommodates the tunnel portal, is designed for serving nature values.

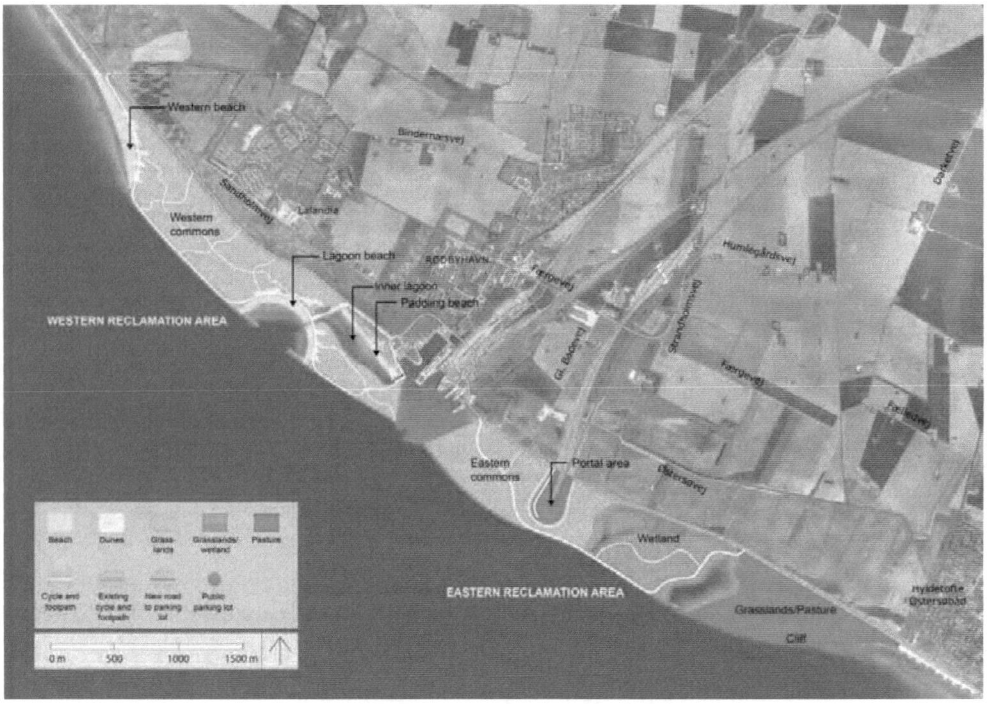

Figure 7. Design proposal for a land reclamation area on Lolland in Denmark

These new features will introduce positive aspects by supplying new natural, environmental and recreational values and will partially rehabilitate an area that has suffered severely from past engineering flood protection projects. The use of this surplus sediment will allow the re-creation of features such as those discussed below.

5.2 The landscape

The new landscapes will connect the tunnel portals to the adjacent coastal areas. This will be achieved in a gradual and harmonious way, thus minimising the visual aspects via a green transition zone. On the German side the natural elevation forms a kind of hill, which on one hand will hide the portal structure from the hinterland, and on the other hand secure the tunnel portal against raising sea level and minimizing the scour protection mound. In contrast to a more traditional design, the elevated landscape would probably have been removed and replaced by a rubble revetment mound. Finally, the passengers are allowed to overlook their journey across the sea.

Figure 8. Tunnel portal on Fehmarn. The elevated landscape was retained. No larger revetment was necessary like on the Danish side. (see Figure 9)

5.3 Protected areas

The existing dike on Lolland, which runs along the coast line, is not influenced by the new reclamation areas in front and will consequently still protect the low lying hinterland. The reclaimed areas will be designed with different perimeter types dependent upon the different technical and environmental functions.

Figure 9. View of the portal area at Lolland protected by a revetment

The stretch around the portal at Lolland will be scour protected. The structures are designed to minimise the visual aspects of the tunnel portal but also to protect against possible rising sea levels due to climate change.

5.4 Re-establish previous landscape

Some of the environmental values were lost in connection with the construction of a major dike along the coast in the area following a major flooding event in 1872. The dike cut off the previous shallow archipelago, Rødby Fjord and other shallow areas, which were later reclaimed by the installation of a pumping station. The historical landscape with Rødby Fjord is seen in Figure 10, along with the reclaimed areas

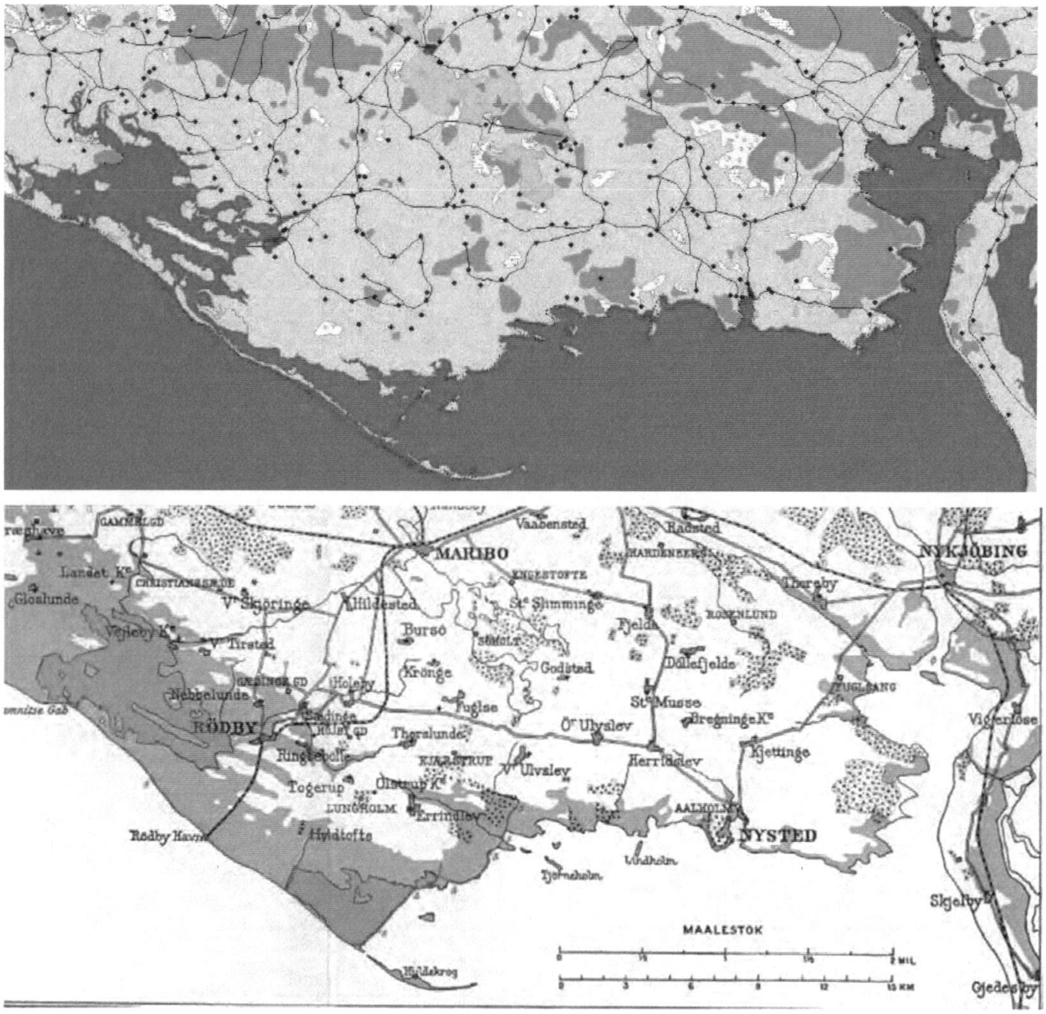

Figure 10. Upper; Historical map for the south coast of Lolland produced by the Scientific Society 1763-1805 – note the large Rødby Fjord. Lower: present condition, the dike (red line), flooded area during the storm event in 1872 (blue area) and dry areas (green). The dike is now protecting the hinterland for flooding and secures with its pumping station and drainage system dry areas for agriculture. With the land reclamation new wetlands are created re-establishing some of the lost previous high environmental values.

5.5 Artificial lagoon

Wetlands, salt meadows and grassland lost due to the construction of the new dike will, to a certain extent, be re-established at the new land reclamation area east of Rødbyhavn, see Figure 11. Overall it will be a 3.7 km long and 0.5 km wide green band featuring an artificial lagoon with two fixed openings east of the tunnel portal. The lagoon includes wetlands, a major recreational island and a small sea bird island. The vegetation in the nature and wetland areas will be allowed to develop naturally, which will enhance the biodiversity of this environment

Figure 11. Visualisation of wetland in the eastern part of the land reclamation area in Denmark on Lolland. The bird island is the small island in the background of the lagoon

5.6 Cliffs

The ferry harbour, Rødbyhavn, was constructed over a century ago and extends 500 meters out from the coast line. This extension has resulted in a beach with sand dunes that has built up west of Rødbyhavn but, on the eastern side of the harbour, the blocking of the littoral transport has resulted in sand erosion along approx. 3.5 km of the coastline, see Figure 12, left.

Figure 12. Left; Protected dike east of the harbour, where beach and environmental values have been lost. Right; Cross section of the cliff section

Most of the eastern area will be filled up to a level of 7m. Natural erosion is allowed here whereby a "natural" cliff will form and the eroded sand will be transported eastward by the predominant littoral transport. This supply of sand will help to stabilise the beaches to the east, following a smooth transition area.

5.7 Artificial beaches

On the western side of the harbour, the area with its beaches will be used for recreational and leisure activities, which feature a major leisure area. The recreational and leisure value will increase with the construction of the beaches and grasslands. Three artificial beaches are planned; on the Danish side, one at the extreme west end of Lolland and one lagoon beach in the middle plus a paddling beach close to the harbour. On the German side a grassland and a beach east of Puttgarden harbour (Figure 8) are planned. The beaches are designed in their equilibrium orientation by fixed structures;

Figure 13. Cross section of the new western beach section with dunes at Lolland

5.8 Reefs

Stones and boulders on the seabed form hard substrate to which sea algae will attach and start to increase the biodiversity, which gradually leads to reef formations. Many of these stones and boulders have been removed for the purpose of constructing harbours, piers, revetments etc. over the last 100 years or more. Stones and boulders large enough to create artificial reef structures will be used for the protection layer on top of the tunnel close to the coast – in shallow water where there will be no barrier effect on the water exchange. If feasible surplus "reef-stone" from the dredging work will be placed at the Natura 2000 area, Sagas Banke, in order to mitigate earlier stone removal over a 25 ha area, see Figure 14.

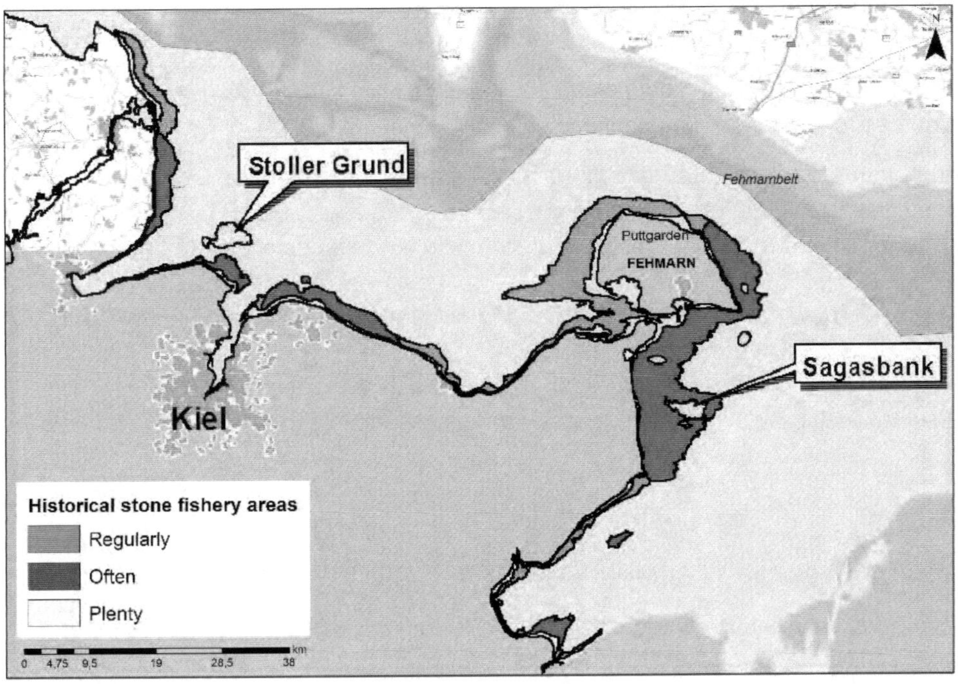

Figure 14. Sagasbank, Germany has suffered for intensive stone fishery and 25 ha will be re-establish

6. STAKEHOLDER ENGAGEMENT TO IDENTIFY POSSIBLE WIN-WIN OPPORTUNITIES

Extensive stakeholder engagement was obtained for this project already from the start. The public, professionals and NGO organisations, as well as authorities were invited to comment on the project. An exhaustive range of consultations (some public) was conducted along the way. This included the scoping process and environmental impact assessments. There were also public plan application document exhibitions.

Cross-border project consultations were conducted in both Germany and Denmark, as well as one involving all the countries around the Baltic Sea (ESPOO hearing).

The outcome of this stakeholder input has had a major influence on the whole project, including the marine area and the new reclamation landscapes; the marine ecosystem's functionality; the bird migration routes; and, of course, the Fehmarnbelt as a transport corridor for other marine life between the North Sea and the Baltic Sea. They are all part of the Danish plan approval already granted and the German plan approval expected in 2018.

7. CONCLUSIONS

This is an outstanding example of the application of the Working with Nature principles for large marine infrastructure projects. The construction of the Fehmarnbelt project is planned to start in 2020. The main features of the project can be summarised as:

- The proposed marine and landscapings are a win-win situation discussed and proposed during the stakeholders engagement
- 19 million m^3 of material will be used to create new landscapes instead of marine dumping
- These will add new nature, environmental and recreational services much to the public needs and requirements
- Understanding and mimicking nature play a key role in the project vision
- Immersed tunnel avoids long-term over-water disturbance of the aquatic environment

8. REFERENCES

CEDA Information Paper, ENVIRONMENTAL MONITORING PROCEDURES, April 2015

Mangor, K. 2011. Designet Natur – fortællingen om et nyt kystlandskab på Lolland og andre kunstige kystmiljøer. Wiljelm +10 Natur i Danmark Conference, Copenhagen, Denmark, 18 November 2011

PIANC Position Paper. Working with Nature, October 2008; revised January 2011.

A guide for applying working with nature to navigation Infrastructure projects. PIANC Workgroup EnviCom 176

https://femern.com/en/Construction-work/Production-site-at-Rodbyhavn/Newrecreational-coastal-areas-to-evolve

PORT OF OAKLAND'S VISION 2000 MIDDLE HARBOR BASIN PROJECTS, OAKLAND, CA, USA

by

Ellen Johnck

ABSTRACT

The Port of Oakland is the primary and only bulkhead container port for the San Francisco Bay region and northern California. It is the fifth largest in the United States based on TEU cargo volume.

The purpose of this paper is to describe the Port of Oakland's Middle Harbor Basin Projects as an example of how a port can work with nature to achieve its navigation efficiency and capability enhancement goals in concert with its environmental and community goals.

The MHB Projects are also discussed as a case study in the PIANC Working with Nature (WwN) Work Group #176 Report currently being finalized for publication scheduled in 2018. ***

The MHB Projects evolved with the Port of Oakland's Year 2000 Vision Plan. The Plan was a $1.2 billion capital expansion plan to build a new modern port for the 21st century. It included several projects to fulfill both the Port's navigation mission by modernizing its terminals and deepening the Inner Harbor shipping channel; and its community and environmental stewardship goals by creating a public park space and shallow water habitat.

The Vision 2000 navigation projects included the dredging of the federal channel from -42' to -50'; building two new marine terminals, a joint inter-modal rail terminal, realignment of roadways, and conversion of the closed Oakland Army Base to maritime use.

The Vision 2000 community and environmental projects were: 1) a 38-acre Middle Harbor Shoreline Park (MHSP); and 2) a 181-acre Middle Harbor Enhancement Area (MHEA): a shallow water fish and wildlife habitat area to be created by beneficially reusing the approximately, 5.2 MCY of the 12-14 MCY of sand and mud from the proposed -50' channel deepening project.

This paper explains the process of how the Port followed the basic WwN template in the implementation of its Vision 2000 projects: **Step I: Establish project goals and objectives; Step 2: Understand the Environment; Step 3: Make meaningful use of stakeholder engagement; Step 4: Prepare project proposal/design to benefit navigation and nature; Step 5: Build/Implement; Step 6: Monitor, adapt, manage**

The Vision 2000 plan called for a new modern port to ensure the Port of Oakland's future economic and environmental success. This meant expanding the Port's shipping operational capacity and improving its cargo handling efficiency while simultaneously achieving its environmental and local community recreational goals.

***__NOTE:__ PIANC's Environmental Commission (EnviCom) published a position paper, first in 2008 and later revised in 2011, that describes the Working with Nature (WwN) philosophy. The position paper defined the WwN concept and discussed, in general, how the approach can be applied to navigation and port infrastructure. The WwN approach provides a basis for maximizing opportunities for working with natural processes to deliver environmental restoration results that go beyond just avoiding or compensating environmental impacts. The Position Paper did not discuss specific tools and practices, Hence EnviCom established Working Group #176 to write a guide for WwN applicaton. The Report No. 176-2018 Guide for Applying Working with Nature to Navigation Infrastructure Projects will be published later in 2018. The following is a case study from the Report illustrating WwN principles at the Port of Oakland,

Ellen Joslin Johnck, RPA, ellen@ellenjohnckconsulting.com, San Francisco, CA. USA

MHB PROJECTS OVERVIEW

The MHB projects were developed as part of the Port of Oakland's (Figure 5.3-1) Year 2000 Vision Plan. Vision 2000 was a $1.2 billion capital expansion plan to build a new modern port for the 21st century. It included several projects to fulfil both the Port's navigation mission, by modernizing its terminals and deepening the Inner Harbour, and its community and environmental stewardship goals by creating a public park space and shallow water habitat area.

Vision 2000 included the dredging of the federal channel from -42 to -50 feet (-12 to -15 metre) deep, jointly cost-shared 50/50 between the port and the USACE; and the building of two new marine terminals, a joint inter-modal rail terminal, and the realignment of roadways

Figure 5.3-1. Port of Oakland looking west

The Vision 2000 community and environmental projects included a 38-acre (15-hectare) Shoreline Park, and a 181-acre (73-hectare) shallow water habitat area to be created by beneficially using the sediment from the proposed -50 ft. (-15 m) channel deepening project (Figure 5.3-2). The following narrative describes how the port followed the WwN six-steps process in the implementation of the Vision 2000 MHB Projects. .

Figure 5.3-2. Vision 2000 MHSP and MHEA Projects.

STEP 1: ESTABLISH PROJECT NEEDS AND OBJECTIVES

The Port of Oakland is the primary and only bulkhead container port for the San Francisco Bay region and northern California. It is the fifth largest port in the US. Vision 2000's call for a modern port that ensures the Port of Oakland's future economic and environmental successes meant expanding the Port's shipping operational capacity and improving its cargo handling efficiency to keep up with the growth of container traffic and size of ships.

In 1998, the Port acquired the US Naval Supply Center, a 500-acre (200-hectare) site adjacent to the Middle Harbor Basin (Figure 5.3-3).

The Vision 2000 strategy was to redirect the navigation use for the US Navy's maritime operations away from the Middle Harbor Basin to the Port's Inner Harbor, the locus of the majority of its terminals. This redirection of the Basin's navigation function would open the Basin and shoreline space for implementing the Port's community and environmental restoration goals.

A commensurate objective was to establish a site for dredged material placement and beneficial use for the 12-14 million cubic yards (9.2-10.7 million cubic metres) of dredged material from the Inner Harbor channel-deepening project. The proximity of Middle Harbor, adjacent to the Inner Harbor, provided a cost-effective solution as a placement site for a portion of the dredged material from the proposed deepening project, the specific amount to be determined. Port of Oakland and U. S. Army Corps of Engineers (USACE) (1998)

Figure 5.3-3. 1999 Oakland Naval Supply Centre shortly after transfer from US Navy to port

The primary community goals for the shoreline park were to gain physical and visual access to the shoreline for the public, which had been closed by the military since World War II; to create a place for learning about the local history, natural environment, and maritime activities; and to foster stewardship for the environment.

STEP 2: UNDERSTAND THE ENVIRONMENT

The Port of Oakland participated in a San Francisco Bay region environmental study in the joint state-federal program known as the *San Francisco Bay Long Term Management Strategy (LTMS) for Dredged Material Placement* (1990-2002). The LTMS's goal was to establish a new environmental dredging plan for San Francisco Bay and an improved and coordinated permit process. (USACE et al, 2001)

The LTMS conducted comprehensive studies of the navigation and biological environment using sediment transport, hydrographic, and biological surveys. The cornerstone of the LTMS was the principle that dredged material is a resource, not a waste, and should be used beneficially to restore tidal marshes, wetlands, levees, and other infrastructure features. The LTMS dredging study, its findings, and the resulting LTMS Management Plan served as the environmental framework for the Port's Vision 2000, including the design and construction of the Middle Harbor Shoreline Park (MHSP) and Middle Harbor Enhancement Area (MHEA) projects.

In addition to the LTMS Management Plan, the Port created a Habitat Technical Advisory Committee (TAC) whose members included state and federal wildlife biologists, including eelgrass specialists, and members of the local community. The goal of the TAC was to develop more detailed environmental information and a site-specific design for the MHEA site. The proposed MHEA restoration target was to be a mosaic of shallow water habitats including sand and mudflats, eelgrass habitat, bird roosting areas, and a small marsh. It was to be implemented and managed through the application of adaptive management principles which were formulated in the *Construction Period and Long-term Monitoring, Maintenance and Adaptive Management Program* adopted by the TAC in 2001. (Winzler & Kelly, Merkel & Associates 2001-2002)

A key restoration engineering challenge was establishing suitable elevation and sandy material conditions for growing eelgrass in the transition from a deep-water navigation basin to a shallow water habitat. An uncertainty was whether the Oakland Inner Harbour dredged material, which is comprised primarily of silt, clay, and some sand, would be sufficiently stable to remain in place to maintain target elevations for each of the various targeted habitats—eelgrass, bird roosting islands, and the marsh.

STEP 3: MAKE MEANINGFULUSE OF STAKEHOLDER ENGAGEMENT TO IDENTIFY POSSIBLE WIN-WIN OPPORTUNITIES

Stakeholder involvement for MHEA began with the Port's participation in the LTMS stakeholder group during development of the LTMS Management Plan (1990-2001) and its implementation, which continues to date, and through engagement with the TAC.

The planning and design of MHSP was done through extensive community involvement in multiple meetings with residents of Oakland and nearby communities in Alameda County. The Port's active engagement in the LTMS stakeholder workgroup coupled with the stakeholder TAC of agency and environmental experts was critically important for moving the Vision 2000 projects forward. MHSP and MHEA were envisioned as win-win solutions, benefiting navigation, the environment, and the community.

The LTMS planning process, including the development of the TAC adaptive management program, was instrumental in achieving public support and the issuance of federal and state permit authorisations for the Vision 2000 projects: the construction of the new marine terminal in the Inner Harbour, dredging of the channel entering the Port to a depth of -50 feet (-15 metres), and construction of MHSP and the MHEA projects. The permit authorisations occurred almost simultaneously with the adoption of the LTMS Plan's dredge-material placement strategy of 40%

upland placement; 20% ocean placement; and 20% in-Bay placement, thereby establishing a new direction for dredged material management through beneficial use.

STEP 4: PREPARE INITIAL PROJECT PROPOSAL/DESIGN TO BENEFIT NAVIGIATION AND NATURE

Based on comprehensive environmental investigations and stakeholder involvement, the Port finalized its dual MHB Projects' designs, which were approved unanimously by the state and federal government permits in 2003. The following features were included in the designs:

- A 38-acre (15-hectare) MHSP, providing public access to the shoreline and the Bay with benefits for nature education and beach recreation (Figure 5.3-4).
- A 181-acre (73-hectare) beneficial use MHEA Project providing shallow water and eelgrass habitat, created with approximately 4-6 million cubic yards (3.1-4.6 million cubic metres) of sediment (Figure 5.3-5) from the Oakland Harbor channel deepening.
- A 5-acre (2-hectare) demonstration marsh; bird island(s) for roosting; and a 10-year habitat performance evaluation.
- The regulatory agencies supported the concept of the co-existence of the proposed MHSP and the MHEA as an ecological reserve of shallow bay and shoreline habitats with its commensurate goal as a beneficial reuse site for the Inner Harbour deepening. This habitat would be important for many species, such as Dungeness crab, flatfish, anchovy, herring, and perch and would serve the purpose as an environmental educational laboratory for the study of habitat restoration and marine biology.

Figure 5-3.4. MHEA and MHSP today

STEP 5: BUILD AND IMPLEMENT

MHSP was completed in 2006-07. Construction of MHEA took place in 2003-2007, concurrent with the commencement of dredging operations for the Inner Harbour navigation channel. The Inner Harbour deepening to -50 feet (-15 metres) was completed in 2008, and almost 6 million cubic yards (4.6 million cubic metres) of dredged material was placed in the MHEA. The US Army Corps of Engineers completed a geotechnical study in 2010/2011 to verify the MHEA's geotechnical conditions. The study's conclusion was that the building/establishment of the MHEA was incomplete.

During monitoring and adaptation discussed in Step 6, the issues for the incomplete construction and establishment of the MHEA were addressed.

Figure 5-3.5. Middle Harbor bird roosting islands

STEP 6: MONITOR, EVALUATE AND ADAPT

The greatest uncertainty to achieving the desired habitat objectives in the MHEA resided in the design predictions that were made regarding the sediment consolidation rates and hydrodynamics. The Habitat Design Criteria Model agreement prepared by the TAC in 2003 contained habitat elevation guidance for the proposed timing of the eel grass planting. The main design concept allowed the dredged material to settle and consolidate after placement, letting nature take its course to foster plant and benthic community growth.

Ongoing monitoring during and after the 2003-2007 channel dredging and dredged material placement phase identified physical constraints within the Middle Harbor sediment transport system that presented challenges to constructing the wetlands portion of the project. In short, the dredged material was settling without sufficient consolidation to enable the eelgrass planting. The USACE geotechnical study referenced in Step 5 concluded that additional material would be needed to further the restoration objectives and recommended a revised sand and mud redistribution plan.

Following the recommendations in the revised sand and mud redistribution plan, approximately 400,000 cubic yards (300,000 cubic metres) of sand was redistributed within the site to further stabilise the sediments.

In 2014, the Port and USACE returned to the regulatory agencies and presented their findings that the creation of the shallow water habitat area was taking longer than originally planned and permitted in 2003, and that they would reconvene the TAC and refine the adaptive management program. (BCDC Staff Report and Meeting Minutes 2014 and 2015)

In February 2017, the TAC concluded that more regular meetings and presentations of monitoring results were needed. An unanticipated benefit during the many years of construction monitoring has been that the Middle Harbor site is proving to provide benefits to address sea level rise and flood protection, and may serve as a demonstration project to address sea level rise in other Bay Area locations. (Port of Oakland 2017)

In 2017, the TAC further concluded that the MHEA site is now stabilised, and interim shallow water habitat has been created. The roosting bird islands and a demonstration marsh have been created. Phase 1 of the eelgrass planting is planned for 2019 with a second phase scheduled for 2020. It is anticipated that the 10-year habitat performance evaluation will begin in 2019.

This case study highlights a universal challenge for ports and the maritime industry to succeed on multi-lateral fronts in the 21st century and illustrates the path to success by working with nature (WwN).

References

Middle Harbor Enhancement Area, Construction Period and Long-term Monitoring, Maintenance and Adaptive Management Program. (2001) (revised 2002). Prepared for U.S. Army Corps of Engineers, San Francisco District and Port of Oakland by Winzler & Kelly, Consulting Engineers and Merkel & Associates, Inc. November and May. Available online at https://www.portofoakland.com/community/environmental-stewardship/habitat/

Port of Oakland (2017) Middle Harbor Enhancement Area Technical Advisory Committee Meeting Summary. February 27. Available online at http://portofoakland.com/environmentalrestoration

Port of Oakland and U.S. Army Corps of Engineers (1998) Oakland Harbor Navigation Improvement (-50 Foot) Project. Volume 1: Final Feasibility Study. Sch No. 97072051. Oakland, CA, USA. May. Available world wide web, public library commons.

San Francisco Bay Conservation and Development Commission (BCDC) (2014) Staff Report on Middle Harbor Enhancement Project, October. Available online at: http:// www.bcdc.ca.gov

San Francisco BCDC Meeting Minutes (2014) re Briefing on Middle Harbor Enhancement Area and BCDC Consistency Determination No. C000.014.00 and BCDC Permit No. C000.014.00 for the

Oakland Harbor Navigation Improvement Project. November 6. Available online at http:// www.bcdc.ca.gov

San Francisco BCDC Commission Meeting (2015) Staff Presentation on Middle Harbor Enhancement Area. Available online http://www.bcdc.ca.gov/cm/2015/0716MHEAPresentationFial2015714.pdf. July 16.

U. S. Army Corps of Engineers (USACE) et al. (2001) (USACE, USEPA, BCDC, SFRWQCB). 2001. Long Term Management Strategy (LTMS) for the Placement of Dredged Material in the San Francisco Bay Region Management Final Plan.. Available online at: http://www.spn.usace.army.mil/Missions/Dredging-Work Permits/LTMS/LTMS Program. July.

UNSUITABLE FILL MATERIAL MANAGEMENT IN PORT TERMINAL CONSTRUCTION: EXAMPLE IN BUENAVENTURA PORT, COLOMBIA

by

Jordan Lagnado[1]

ABSTRACT

The construction of COMPAS' Bulk terminal in Buenaventura Port, on the Pacific coast of Colombia took place in the Aguadulce Peninsula which is a narrow piece of land surrounded by sea. Soletanche Bachy International, in joint venture with Soletanche Bachy Cimas and Constructora Conconcreto were in charge of the design and build contract for the main infrastructures, mainly: the offshore 250m long wharf on piles linked to the land by a 200m long trestle, the conformation of yards and access road by earth movement, pavement and multiple civil structures.

Based on the available ground investigation reports, the design for the yards already considered to use the cut material above the yard final elevation to fill in the lowest parts. The cut material was mostly clay and after lime-stabilization process, it could meet the physical properties required for fill. However, all excavated material was not suitable for backfilling purpose due to its organic content, high water ratio and plasticity index. Therefore, the lack of space to dispose large volumes of unsuitable material inside the terminal boundaries was a critical point during the construction phase as this volume increased significantly. After some unsuccessful attempts to find a suitable dosage of lime, cement or polymer, project team had to find the solution in the disposal areas design. Indeed they built strong vertical retention structure to come as close as possible to the statutory boundaries of the terminal to fit the surplus volume. Some retaining walls were made of sheet pile, others with geotextile megabags filled with the same cut material. Inside the deposit area, due to the high seismicity in the region, it was also required to use ground improvement techniques to guarantee the stability of the bulkhead. Rigid inclusions or consolidation (wick drains with surcharge) were the two techniques employed, and finally extended to the whole stockpiling areas to reach at the final yard level a bearing capacity of 2 ton/m2 and 5t/m2 in order to deliver to the final user an additional yard surface ready for a short-term port development.

That is a good reference for a win-win project where design and site constraints are evolving in line with owner's needs.

INTRODUCTION

The city of Buenaventura on the Pacific coast of Colombia hosts the most important industrial port of for the economy of the country. Due to its favorable location facing Asia, the port is not only a strategical hub for trade with China but also with all countries on the Pacific coast. This explains why in front of the historical port, on the narrow Aguadulce Peninsula, two new terminals have been built under a joint development: COMPAS grain and coal terminal, and SPIA container terminal. Figure 1 shows the initial shape of the peninsular land and the final drone picture.

The COMPAS terminal has been designed to service bulk carriers of 80,000mt, and its onshore infrastructure can store general cargo plus 54,000mt of imported grain and 110,000mt of coal for exportation. It was built in the period 2015 to 2017. The project involved 1) construction of the offshore 250m long wharf on piles linked to the land by a 200m long trestle; 2) the earth movement for the land reclamation; and 3) bulk handling yards and access road, road pavement and multiple civil structures. The project was executed under a design and build contract by the joint venture Soletanche Bachy International (SBI), Soletanche Bachy Cimas (SBC) and Constructora Conconcreto (CCC). BergerABAM Inc. (BA) was the design company chosen by the joint venture.

[1]Soletanche Bachy International, Major projects department, jordan.lagnado@soletanche-bachy.com, France

Figure 1: Peninsula Aguadulce before and after

PROJECT GENERAL DESCRIPTION

Offshore structures: wharf and trestle

The offshore structure is a concrete open quay, founded on 233 steel piles (diameters 1067mm and 914mm) driven into the seabed. Dimensions of the wharf are 250m long, 32m wide until the extreme end where it's increased to 40m to allow the trucks to make a U-turn. Access to the quay is through a trestle of 190m by 14m wide available to vehicles and walkway, with additional 4m-cantilever beams for the double deck conveyors. Figure 2 shows the wharf and trestle under construction.

In order to shorten the construction time, reduce the safety risks and environmental impact, the design-build team looked for simplification of the tasks done offshore, and particularly to reduce the volume of in situ concreting. Therefore, the use of precast concrete elements has been encouraged and as a result, the quay was delivered to the owner 5 months ahead of the contractual schedule. Allowing him to start earlier the operation of the terminal and generating the financial benefits that come naturally with.

Moreover, with this alternative design the resulting structure had an optimal "seismic" mass and that was a big advantage given the earthquake vulnerability in the area.

Figure 2: Wharf and trestle under construction. Progress 80%

Initial Terminal yard layout and boundaries

On the shore side, the terminal is composed of the following principal structures: four vertical silos of 28,000mt total capacity, two horizontal silos of 26,000mt total capacity, and a coal yard of 110,000mt. The platforms are situated around the elevation +5,30MSL with a total area of approximately 6.5Ha.

A dedicated access road running along the North side of the peninsula connects the terminal yards to the main Aguadulce Port entrance and is independent from the adjacent SPIA Container Terminal.

The layout implied reclamation works with the objective to double the native land surface inside the concession area mainly using the same excavated material (Figure 3). On the North side of the terminal, a disposal area was designated to store the unsuitable fill material, and let it self-consolidate for a few years as no immediate development was envisaged at this early stage.

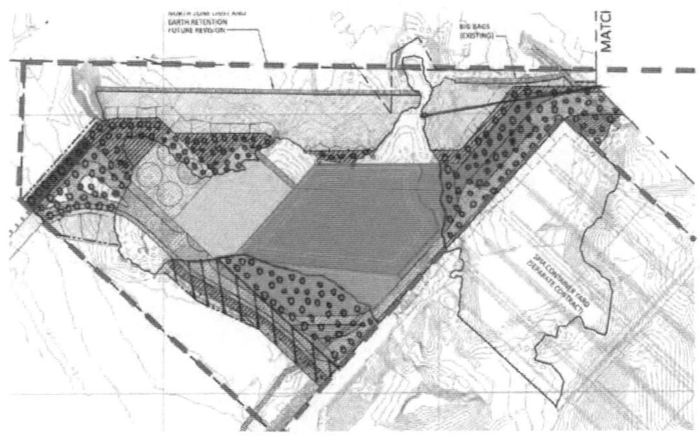

Figure 3: Cut and fill areas at project start date (access road is not shown)
(fill shown with handmade hatch, red discontinued line is concession boundary, pink area is initial unsuitable fill material disposal area)

Design assumptions and initial design for reclamation areas
The design involved reutilizing the in situ excavated material above the elevation +5,30MSL to backfill the lowest spots and to obtain a levelled platform with the required bearing capacities at +5,30MSL. The excavated material was identified mainly as altered claystone and after a stabilization process with lime mixing (around 4% of lime were used), it met the required minimum UCS (Unconfined Compressive Strength) at 7 days of 200KPa and 300KPa for backfilling.
The design-build team had to conceptualize measures to prevent potential of flooding of the reclamation areas by tidal action (mean tide amplitude 4,50m) which occurs almost half day long. The backfilling works in most areas occurred behind a permanent sheetpile wall that could play the role of retaining structure and shore protection at the same time. In some places with shallow retaining height, reclamation occurred behind a temporary dike made of geotextile bags filled with excavated material. Refer to Figure 4 for the cross sections. Parts of the reclaimed areas situated on top of deep saturated silty deposit (upto 10m thick layer) required soil improvement prior to general backfill process. Soil improvement by concrete inclusions was selected due to its fast execution time.

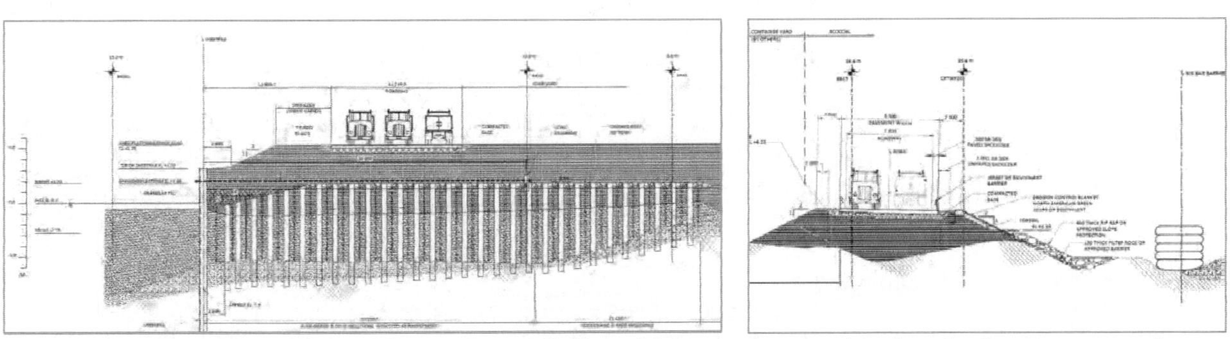

Figure 4: Sheetpile bulkhead for coal yard (left) and temporary megabag barrier for access road (right)

It is interesting to mention also two other environmental conditions that had to be considered.

The first one is the seismic activity. Buenaventura is situated at the toe of the Occidental Cordillera of Colombia, at the convergence of the Nazca plate and the South American continental plate, as shown in Figure 5. The frequency and strength of the earthquakes in the region is very high. As indicated in the Colombian seismic code NSR-10, the design used a seismic coefficient of 0.4g that corresponds to the peak ground acceleration (PGA) of the Contingency Level Event (CLE).

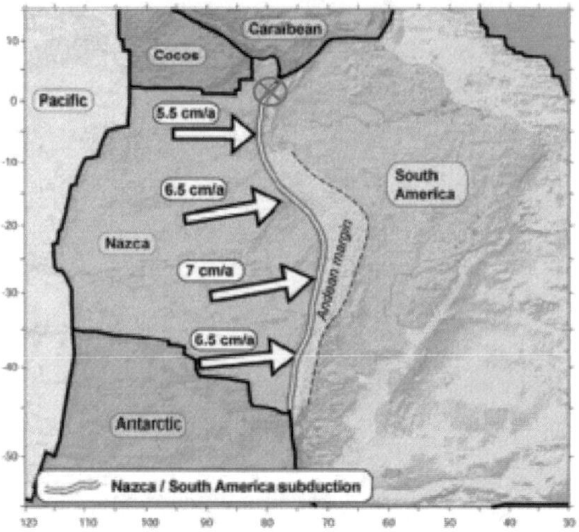

Figure 5: Location of Buenaventura in the south-American subduction system (C.Vigny 2003)

The second local condition is the high quantum of rainfalls and relative humidity. Buenaventura, by its geographical situation mentioned before, is among the cities in our planet with the highest rainfall rates, and does not possess any dry season. The daily average rainfall is 20 liters/m2 (or mm/m2) with peaks reaching 60mm per day, with rainfall mainly occurring at night. Figure 6 shows the monthly rates. The high rainfall was a major consideration in selecting appropriate construction sequence and the methodologies for earth movement. This was true in particular for the lime mixing process and backfilling works that were very sensitive to humidity. For reclamation work, it was not only necessary to block the seawater from entering the area (by megabags barrier or sheetpile wall) but also to drain out the rainwater coming down from the catchment area.

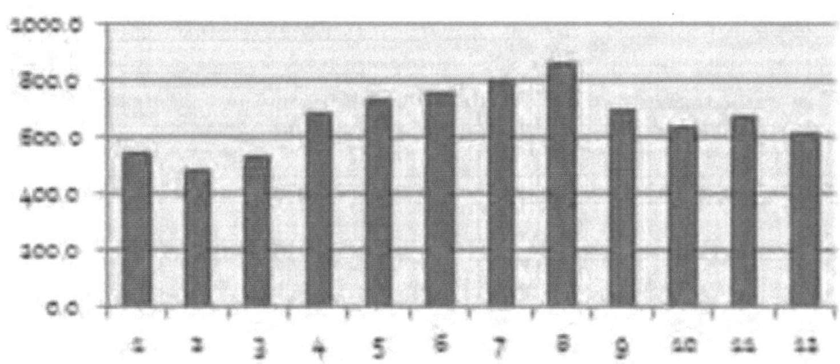

Figure 6: Rainfall histogram in Central Pacific Coast of Colombia (in mm by month)

Challenges with the excavated material

The peninsula in situ material had been characterized by various campaigns of geotechnical investigations during the tender stage (Figure 7 shows the location of boreholes both onshore and offshore). Onshore, the boreholes were executed at the highest topographic elevations around +11.0MSL. The data showed the fractured siltstone covered by layers of soft plastic clay material with high water content, as illustrated in Figure 8. Material description from the boring logs was "(CH) CLAY, yellow and light gray, high plasticity, moist". From the lab tests results, the water content W for these

layers situated in the range 40% to 52% was above the plastic limit WL of 27%, meaning that the soil was saturated.
These layers were initially considered as non-suitable for backfill purpose and their destination was in the North disposal area.

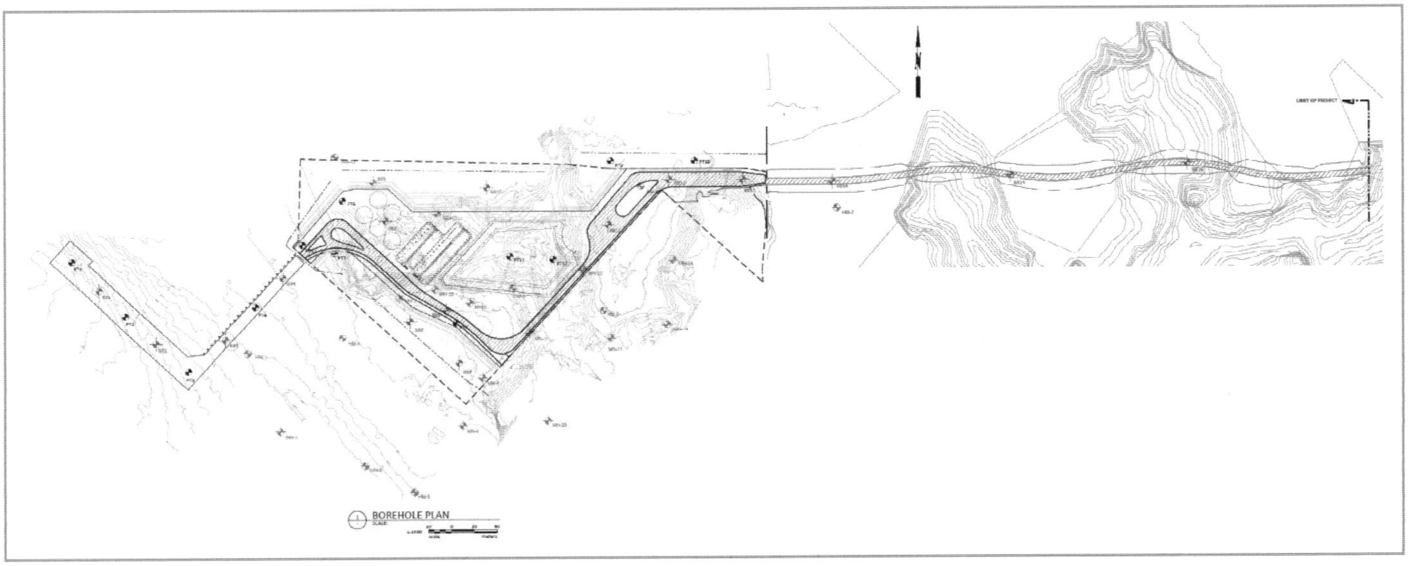

Figure 7: Overall layout of Boscoal terminal and access road, with borehole locations

From the combination of the geotechnical investigation reports and the land survey at tender stage, it was estimated that 290,000m3 had to be excavated to construct the yards and roads of the terminal, including 38,000m3 of unsuitable plastic clay material. However, during the construction phase and the additional work done, the construction team had to deal with a significant increase of that volume. It became a priority topic since the project Environmental license had not considered any disposal area out of the terminal concession boundary. Two options were then investigated.

Figure 8: Yellow color highly plastic clay layer

RESEARCH TO IMPROVE THE SOIL PROPERTIES

The first action of the team was to find a way to use this material in the reclamation process (target 7 days UCS was 300kPa). Further investigation through laboratory tests and site trials were conducted to analyze if other additives would be effective on site and economically interesting. This paper is not focused on that investigation, but the following lines will describe the conclusions. Higher lime dosages

than 4% were tried on site and samples analyzed in laboratory. The UCS results indicated improvement in soil properties for the claystone material but it was not the case for the plastic clay material.
Also, it was studied the effect of mixing with lime 1 or 2% and additional 0,25 to 0,50% of polymers. Lab and small field trials had shown good results, but at a larger scale, this alternative could not reach the same results because of the heavy rainfall that were disturbing the maturation process. It would also require a large infrastructure to mix and store the soil under a shelter on the job site, which was economically not interesting. The same reasons applied to cement mixing.

TERMINAL LAYOUT UPGRADE

The other way to find a solution was to look at the disposal areas themselves.

North yard

It was quickly identified that the planned disposal area on the North side had to be rearranged to accept more material, at least to serve as temporary storage to avoid disrupting the progress of the excavation works. The concept was to expand the area as much as possible and stockpile there all the surplus material. It was made possible due to the owner's foresight since the owner had already set further offshore the concession boundary in view of future development. Therefore, the conditions were created to modify the contour of the project and create a disposal area with the maximum extension that the construction permit could allow. The contour is shown in Figure 9.

However, the owner saw an opportunity to have a new yard of 2Ha in this area; it instructed the design-build team to plan for handling new cargo in the terminal. As a result, the owner then requested to further transform the design-build team to convert the mud disposal area into a yard with a surface load capacity of 5t/m2.

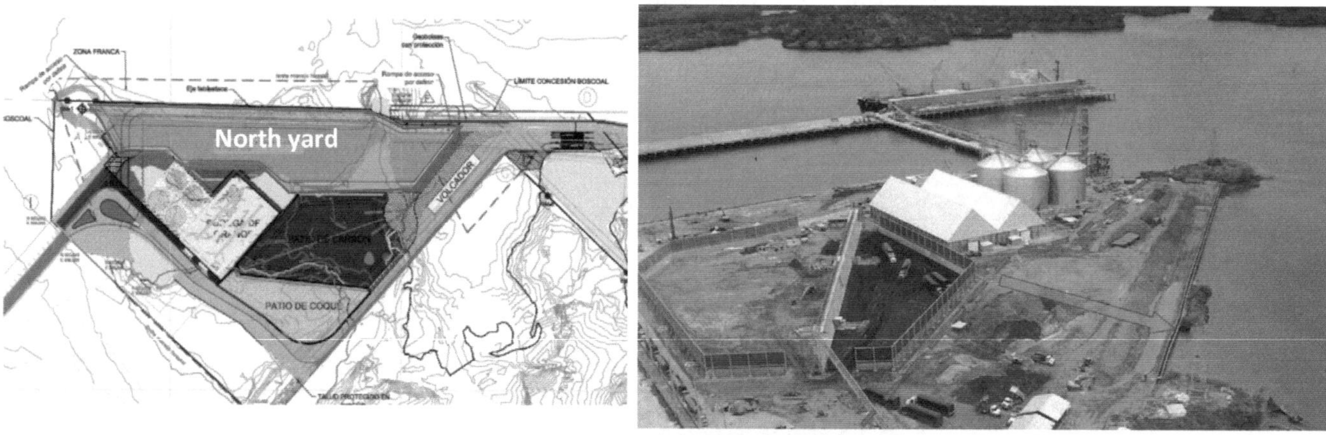

Figure 9: North yard extent

The sequence of work was as follows:
The first task was to build a vertical retaining structure along the concession boundary of 290m long, made of AZ hot-rolled sheetpiles with its wale beam and tie-rods connected to a parallel concrete deadman wall 10m behind. The sheetpiles were installed by both vibratory and impact hydraulic hammers to provide an embedment into the weathered rock located approximately 4m below seabed level. The particularity in that area was the seabed elevation that was above the low tide level (seabed at -1.0MSL, and MLWS at -2.7MSL). A working barge was deployed at the working location during high tide and was purposely grounded on the seabed most of the day; however, it was effective in installing the sheetpiles thanks to its 180t crane with adequate working range.

Once the sheetpile barrier was constructed, the area was filled up and levelled at the elevation +5.30MSL. River granular fill from local quarries was installed to help lower pressure on the sheetpiles and provide proper foundation for the anchor system.

At yard level, an internal road was set up adjacent to the sheetpiles, here again to limit the ground pressure on the wall. In the other parts, to get the final bearing capacity of 2T/m2 at the first stage, and then changed to 5T/m2, the team selected preconsolidation as ground improvement technique above other faster one because of its cheaper cost. Despite the relatively slow process of induced settlement, the selected method satisfied the project schedule.

Figure 10 shows a cross section of the yard where 10,300ml of prefabricated vertical drains (PVD) or 'wick drains' were installed. Then a 2.50m high surcharge was placed to initiate the settlement for a few months. Two types of material were used as surcharge. On the Western portion of the yard, the team used unsuitable fill material. For the Eastern side, as the owner was receiving a vessel with clinker in bulk and was requesting a place to stockpile, the project team naturally proposed to accommodate it in the North yard and utilize it as surcharge. Figure 11 illustrates the settlement vs. time diagram, with a notable difference of the settlement rate using excavated material (density 1.30) and Portland clinker (density 1,51).

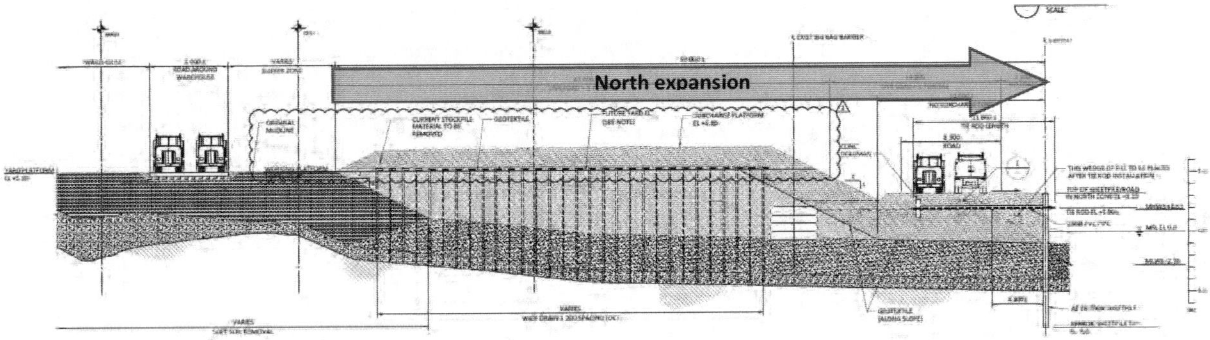

Figure 10: Cross section north yard with typical surcharge (here 1,5m high for 2t/m²)

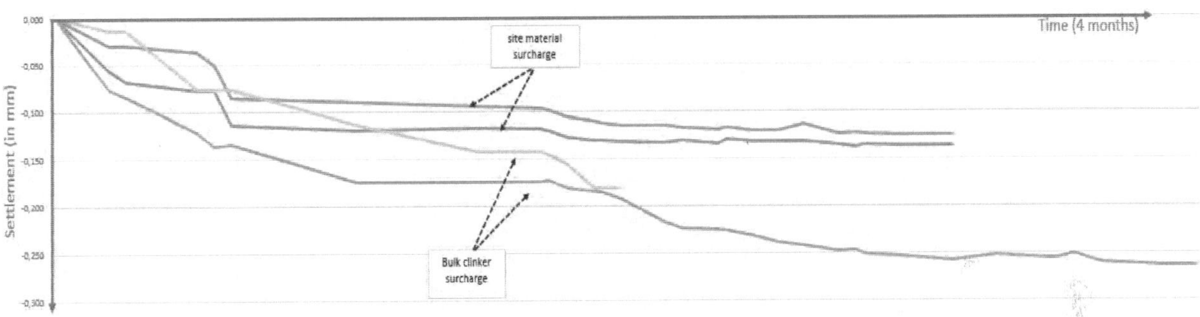

Figure 11: Settlement graphs

South Yard

As the North disposal area had been transformed since an early stage into an operational yard, the area could not bear the same large volume of material that was initially calculated. The owner was open to a proposal for another yard expansion within the allowed boundaries. Therefore, the design-build team developed a solution in the south part of the terminal with a fast track solution to build a 1Ha yard with 5T/m2 bearing capacity. The new reclaimed area is built on the sea side of the sheetpile retaining wall. Figure 12 is showing the aerial view of the area.

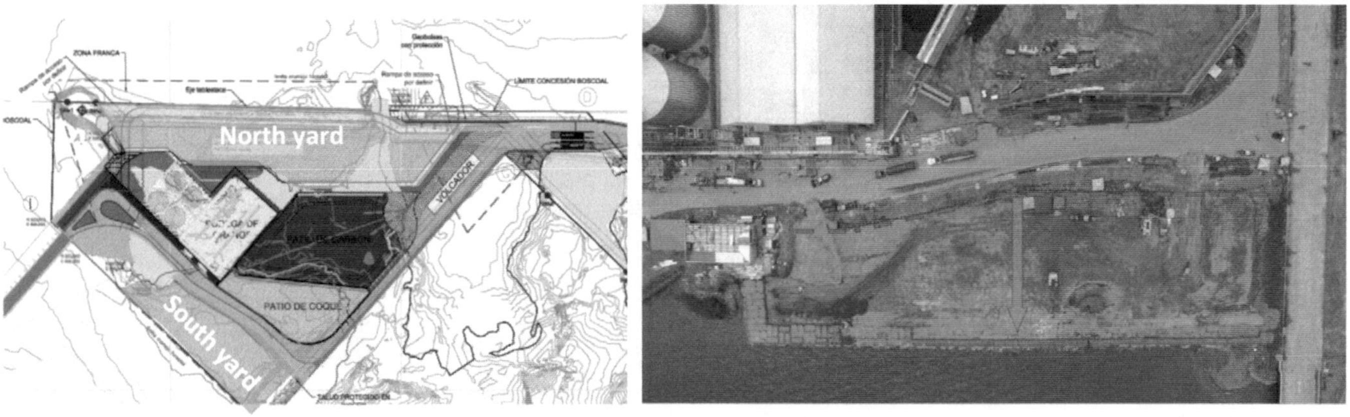

Figure 12: South yard extent

To create this new yard space, fill material was retained by a system of heavy duty geotextile bags filled with the same unsuitable materials directly supported on the seabed and interlocked with each other. A pyramidal dike geometry was formed along the concession boundary with the soil filled bags, as shown in Figure 13. Part of the surplus volume of unsuitable soil was incorporated and confined into the retaining structure, that made this solution adequate to the situation.

Figure 13: Geotextile megabags retaining wall

After the dike was completed (and the lake formed by the rainwater drained out of the confined area), the area was backfilled with the remaining stockpiled material and the surcharge material from the North yard. Above that layer, a working platform was prepared to allow the heavy equipment to access the area and install rigid inclusions through the soft ground (the backfilled material and the marine clay below). 190,000ml of inclusion were installed, mainly with a 2mx2m grid, except along the dike where the grid was changed to 1mx1m with a center rebar in each column to suppress the slip failure risk. See Figure 14 the cross section though the South yard.

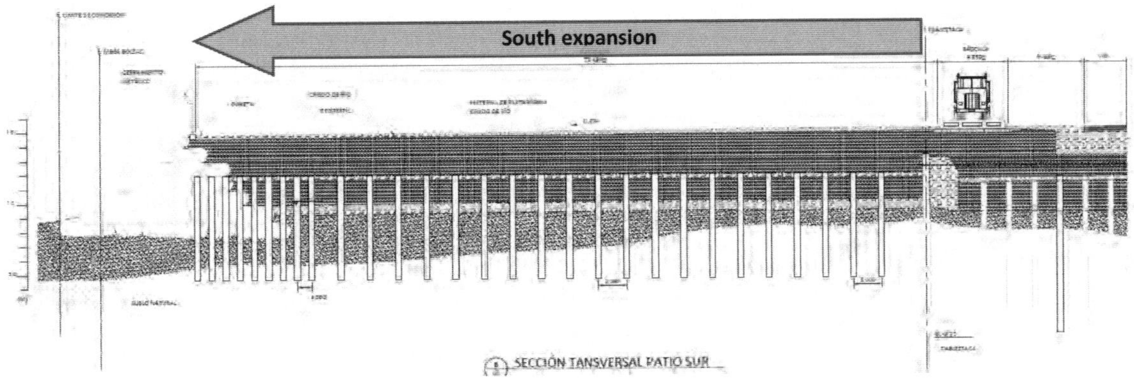

Figure 14: Cross section South yard

Finally, the backfilling continued until the final level 3m above the top of the inclusion to reach +5.30MSL with the remaining lime stabilized claystone available on site and the top layer with of granular fill.

CONCLUSIONS

The construction of the Boscoal Terminal in Aguadulce, Colombia within the tight schedule given the site and environmental constraints reflects a major success due to the interaction between the Owner and design-build team who worked hand in hand to find on time practical and cost-effective solutions to the challenges that appeared during the construction phase. All areas were delivered on time with some ahead of schedule so that the owner could start operation before the full completion of all the works under the scope of the contract. The partnership was moreover witnessed in the use of the unloaded cargo for the purpose of the ground improvement phase, saving natural resources.

Also at the planning stage, it is relevant to highlight that the owner had already planned for the terminal future expansions Northbound and Southbound, and got the statutory permits accordingly. The empty space originally planned to be not developed but available since the first stage of terminal construction was the foundation for the adequate management of the surplus material. The project is a prime example on the importance of conducting detailed and accurate ground investigation campaigns that are necessary to have a detailed characterization of the soil conditions to select an appropriate construction sequence. We think that similar projects in challenging sites should consider this 'win-win' example of partnership between the design-build team members as reference to resolve unforeseen conditions during terminal yard construction.

References

Regionalización de Colombia según la estacionalidad de la precipitación media mensual, a través análisis de componentes principales (ACP) Guzmán D.; Ruíz, J. F.; Cadena M. Grupo de Modelamiento de Tiempo, Clima y Escenarios de Cambio Climático Subdirección de Meteorología – IDEAM (2014)

Reglamento Colombiano de construcción sismo-resistente, NSR-10

Study on Chilean subduction zone, C.Vigny, ENS, France, 2003

Monitoring Dredge Placement Operations through Fine-Scale Suspended Sediment Observations within a Shallow Coastal Embayment

R. Beecroft[1], A. Grinham[1], C. Heatherington[1], Craig Wilson[2] and R. Cossu[1]

ABSTRACT

The design and application of a near-bed monitoring system was an important component for identifying key sediment transport processes within a dredge placement area. Fine temporal and spatial scale current velocity measurements indicated that inequalities in the characteristics of the flood-ebb tidal regime was a contributing factor in the transport of material at the site. As a result, the net transport of suspended sediment was an estimated 610 tons per m^2 at 0.6 m above the bed, oriented in the direction of the principal flood tidal axis. Total suspended solid concentrations were obtained through an array of acoustic and optical instrumentation, providing direct links between the locality of dredge placement operations and increases in near-bed concentration. The application of a vertical suction profiling system enabled concentration estimates from 0.1-0.8 m above the bed, an area particularly complex in nature and difficult to monitor. The study underlines the need to develop near-bed monitoring approaches associated with dredging operations, whereby the majority of the sediment is transport during placement operations or later re-entrained during energetic hydrodynamic periods.

KEYWORDS

Dredge Monitoring, Suspended Sediment, Dredge Disposal

1 INTRODUCTION

Management of dredging operations is a key procedural obligation for both dredge operators and ports around the world. In particular, the assessment of dredge placement area's (DPA's), whereby large quantities of dredged material are relocated and placed at sea. As part of the continual management of dredging operations, so to ensure limited impact on the surrounding environment, intensive environmental monitoring and the development of numerical modelling approaches are tools frequently utilized within the industry. However, the complex nature of both the suspended sediment regime and hydrodynamic conditions governing transport are major contributors to uncertainty surrounding the assessment of dredging operations. Particularly the inability to distinguish between the natural variability of the site and concentration spikes related to the placement of dredge material (Fettweis et al., 2011; Onuf, 1994; Orpin et al., 2004). If insufficient field data within DPA's remains a common trend, how may dredge operators accurately assess the transport of dredge material from within these areas into the surrounding environment?

One of the areas requiring a greater degree of investigation is the transport of suspended sediment in the bottom boundary layer (BBL). The transport of dredge material within this region is typically overlooked, partly due to the logistical difficulties associated with deploying instruments at close proximity to the bed. Elevated suspended sediment concentrations and extreme variability in particle size characteristics also contribute to measurement uncertainty, especially during periods influenced by dredge operations (Agrawal & Traykovski, 2001; Mikkelsen & Pejrup, 2000; Smith et al., 2008). The majority of sediment placed within DPA's is initially transported to the bed as a turbidity current (Gordon, 1974; Johnson & Fong, 1995); hence, it is important to develop robust methodologies to monitor the continual transport of dredge material within this region. By continual transport, this study presents measurements related to the initial dispersion of material during placement operations and the entrainment of unconsolidated dredge material under more energetic hydrodynamic conditions.

Total suspended solid (TSS) concentration measurements presented within this study were obtained during a period of extensive dredge placement operations within Moreton Bay. Traditional measurement turbidity measurements within the BBL were complimented by a vertical array of suction intakes, validating TSS

[1] The University of Queensland, Australia, r.beecroft@uq.net.au
[2] Port of Brisbane Pty Ltd, Australia

concentration estimates under extreme suspended sediment regimes. Section 2 outlines the extensive monitoring approach, highlighting a simplistic and cost effective method to obtain TSS concentration estimates as close as 0.1 m above the bed. The results and analysis of the intensive data series are presented in sections 3 and 4, where transport processes and future recommendations for monitoring approach form key outcomes of the study.

2 INSTRUMENTATION AND METHODOLOGY

2.1 SITE DESCRIPTION

The Mud Island Dredge Placement Area (MI DPA) located within Moreton Bay, Queensland, Australia, is an example of an environmentally sensitive and complex area as it is subject to tides, winds, river run-off and intensive dredging operations. Strategically situated on the western side of Mud Island (Figure 1), the area is easily accessible from the Port of Brisbane. As Australia's largest cargo port, continual requirement for maintenance dredging results in the placement of large quantities of fine silts and clays (d_{50} = 5 μm) within the area. An example of the operational intensity within the MI DPA is provided in Figure 1, with 43 placements conducted during the 7-day monitoring period. Bathymetry at the site is relatively flat, with the bottom-mounted frame deployed 10 m below mean water level. Current velocities and water level fluctuations were driven by a semi-diurnal tidal regime orientated NNW (flood) and SSE (ebb) direction. Dominant local winds are responsible for the small period (T_p <4 secs) wind waves and significant wave heights typically less than 1.2 m. The locality of the Brisbane River was identified as a possible source of sediment to the site during flood events, yet it has no impact on the observations presented within this study.

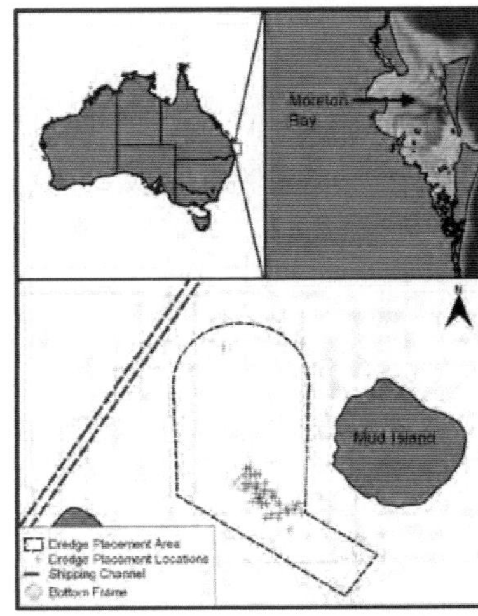

Figure 1: Moreton Bay and Mud Island Dredge Placement Area

2.2 FIELD DEPLOYMENT

Field observations presented within this study were obtained throughout a 7-day period spanning the 8th - 15th February 2018, referred to hereafter as day-of-year (DOY) 38 – 46. The range of instrumentation, respective configurations and elevations above the seabed are outlined in Table 1, with all instruments situated within 1 m of the bed. Three independent methods were employed to estimate TSS concentrations including optical backscatter (*Turner Designs Turbidity Sensor*), acoustic backscatter (*Nortek Vector ADV*) and laser diffraction (*Sequoia Scientific LISST-100X type C*). In addition, a vertical array of 4 mm (inside diameter) suction tubes were also fixed to the bottom mounted frame (Figure 2). The array of tubes were orientated perpendicular to the primary flood-ebb flow direction with the assistance of scientific divers, typically taking 1 minute to fill a 1 L bottle. Physical water samples were pumped to the surface via a constant pressure pumping station located onboard the University of Queensland research vessel and gravimetric filtered through 0.7 µm nominal diameter Whatman GF/F microfiber filters

Instrument	*Configuration*	
Laser In-Situ Scattering and Transmissometry (LISST)-100X	Sampling frequency (Hz)	1
(Sequoia Scientific)	Burst interval (mins)	30
	Samples per burst	120
	Elevation (m)	0.25
Acoustic Doppler Velocimeter (ADV)	Sampling frequency (Hz)	16
(Nortek)	Burst interval (mins)	10
	Samples per burst	30
	Elevation (m)	0.6
HR Acoustic Doppler Current Profiler (ADCP)	Sampling frequency (Hz)	1
(Nortek)	Profiling range (m)	1.05
	Cell size (m)	0.03
Concerto CTD	Sampling frequency (Hz)	6
(RBR Ltd.)	Burst interval (mins)	10
Cyclops 7 Turbidity Probe	Samples per burst	540
(Turner Designs)	Elevation (m)	0.5

Table 1: Instrument Configuration

One of the major advantages of using LISST measurements is the ability to estimate both suspended sediment concentration and distribution of particle sizes in-situ. The LISST relies on small angle forward particle scattering along the 0.05 m measurement cell length, whereby light is scattered onto 32 ring detectors with logarithmically spaced radii at the cell end. Particle shape effects are also considered through a "random shaped" kernel matrix developed by Sequoia Scientific and Agrawal et al. (2008). The random shaped kernel matrix was used to perform the inversion from scattered light intensity on each of the rings to a distribution of volumetric concentration based on the known light scattering properties of these random particles (Andrews et al., 2010). The LISST deployed during this field study recorded estimates for the particle size distribution in the range of 1.9-381 µm. Measurements exceeding the recommended optical transmission (τ) limits ($0.2 < \tau < 0.98$) were also removed from the data series.

The 2 MHz downward facing HR ADCP measured current velocities and directionality within the bottom meter using a pulse-coherent Doppler system. Analysis of the data series indicated that velocity measurements within 0.2 m of the bed were subject to exceedance of the ambiguity velocity as described in detail by Lohrmann and Nylund (2008), resulting in measurements within this region omitted from the analysis. Validation of the HR ADCP velocity magnitude and direction in (bin 12) were also evaluated through cross correlation with the ADV point source velocity measurements situated 0.6 m above the bed ($R^2 = 0.88$). Friction velocity (u_*) estimates were derived from HR ADCP velocities in bins 1-25 (0.2 -1 m above the bed). The procedure assumes the boundary is rough and the flow fully turbulent, thus the near-bed velocity distribution were approximated by the law-of-wall velocity distribution *(1)* in accordance with the *"conventional method"* outlined by Cheng et al. (1999).

$$U(z) = \frac{u_*}{\kappa} \ln\left(\frac{z}{z_0}\right) \quad (1)$$

Profiles not consistent with *(1)* ($R^2 = < 0.90$) were excluded from the analysis, resulting in u* estimates with error of ±12% at the 95% confidence interval. The influence of waves not considered due the short period ($T_p < 4$ *secs*) wind waves and deep-water wave regime.

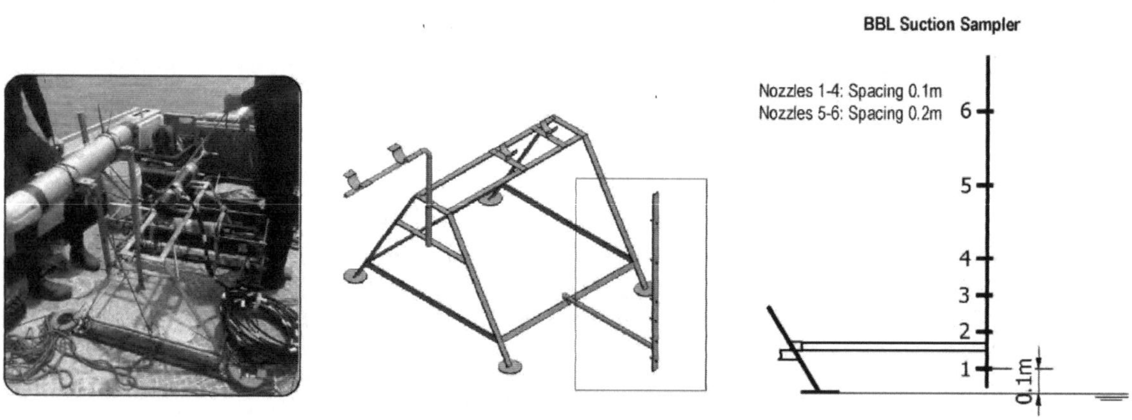

Figure 2: Deployment Frame and BBL Suction Profiler

2.3 INSTRUMENT CALIBRATION

TSS calibrations were conducted for both the LISST and turbidity sensor using sediment obtained from the site. Linear regression between a known TSS concentration, LISST volume concentration (*µl/l*) and turbidity (*NTU*) resulted in multipliers of 1.90 ($R^2 = 0.64$) and 4.07 ($R^2 = 0.75$) respectively. Calibration of ADV acoustic backscatter intensity (BSI) was performed by first analyzing the correlation between BSI and NTU. BSI is defined here,

$$BSI = 1E - 5 * 10^{0.043*Amp} \quad (2)$$

where *Amp* refers to the mean acoustic signal strength (*dB*) across the three acoustic heads (Nikora & Goring, 2002).The calibration process assumes that under the low TSS concentrations observed during the deployment period, the difference in concentration between the ADV ($z = 0.6$ *m*) and turbidity sensor

($z = 0.5\ m$) were small. The correlation between the BSI and NTU for the range of concentration observed during the field deployment was high ($R^2>0.99$), resulting in an estimation for TSS derived from the BSI-NTU relationship in the field.

3 RESULTS

The near-bed velocity regime was governed by mixed semi-diurnal tidal velocity fluctuations, with maximum velocities of 0.35 ms^{-1} most prevalent during flood tides (Figure 3b). Neap tides were observed towards the beginning of the deployment period, resulting in smaller current velocities and little velocity shear close to the bed. As the tidal fluctuations transition towards the spring cycle, current velocities during both the flood and ebb periods increased, with tidal mean u_* estimates increasing by up to 0.015 ms^{-1} during flood tidal periods (Figure 3c). Inequality between the flood-ebb tidal periods were identifiable in the u_* estimates, indicating that the flood tidal period possesses a greater potential for entrainment and transport of sediment in the area. Generally, significant Northerly wind events between days 42 – 45, contributed to a secondary wind driven flow acting constructively with the flood tidal regime. Similar processes have been identified within the upper water column during southerly wind periods that coincided with the ebb tidal regime during deployments conducted in 2017. The data series presented here is the first indication that wind driven flows may be factor within the bottom boundary layer at the MI DPA.

Forty-three dredge placement operations occurred within the MI DPA throughout the deployment period, with the closest located within 100 m of the deployment site. Figure 1 provides a schematic of the placement locations relative to the frame, with majority of the placements located in the south-eastern corner of the DPA. Figure 4a, b presents a direct comparison between the locality of the dredge placements and near-bed TSS estimates derived from the LISST, insinuating the impact of dredge placement operations in the area. Events represented in black refer to placement events upstream of the frame (relative to tidal direction), while those in red refer to events downstream of the frame. From a conceptual standpoint, placement events within close proximity and upstream of the frame should be more likely to register elevated TSS concentrations. The results presented in Figure 4 support this hypothesis, with more pronounced TSS spikes correlated with dredge placements upstream of the frame. Furthermore, the time-series of TSS concentrations underline a clear trend, indicating that the frequency of dredging in the area was associated with larger TSS spikes and an increased standard deviation of the time averaged TSS concentrations. Days 41– 45 for example included both an increased frequency of dredge placement events combined with more energetic hydrodynamic conditions compared to the days prior. As a result, mean TSS concentrations during days 39 – 41 were in the range of 30±16 mgL^{-1}, significantly increasing to 82 ±70 mgL^{-1} during day 41. There are number of instances where TSS concentrations exceeded 250 mgL^{-1} remaining elevated throughout the majority of the more frequent dredge placement period.

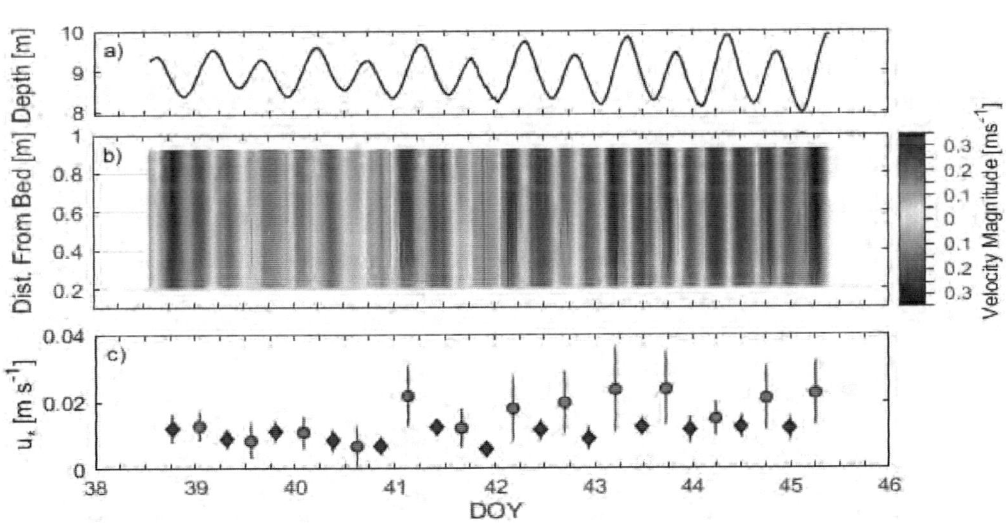

Figure 3: (a) Water Level Fluctuations (b) Near-Bed Horizontal Velocity Magnitude (North-East) and (c) Friction Velocity Estimates

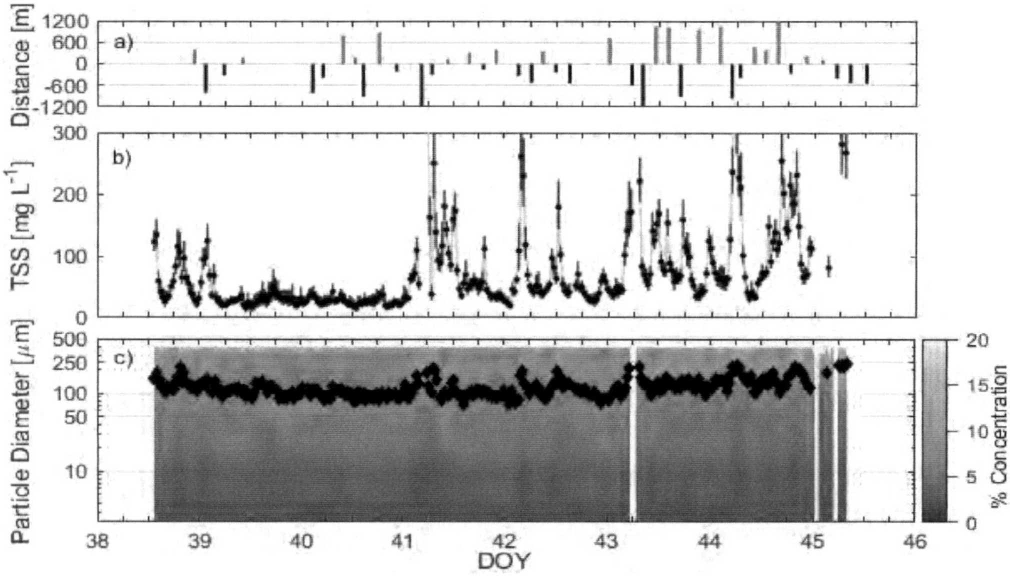

Figure 5: a) Location of Dredge Placements Relative to Frame (b) LISST TSS Concentration Time-Series (c) LISST Particle Size Distribution and d_{50} Estimates

Figure 4c visualizes the distribution of TSS concentration across the measurement range of the LISST (1.9-381 µm) as a percentage of the total TSS concentration. Black markers plotted over the percentage concentration distribution indicate the mean particle diameter (d_{50}). A distinguishable feature of the particle size results is the direct correlation of increasing d_{50} with increasing TSS concentration, especially during the period after day 41. During some of these instances, d_{50} is as much as 150 µm larger than those observed during the lower concentration periods (DOY 39-41). The d_{50} value is a good first order estimate for describing the distribution of suspended particle sizes, yet it may not be entirely representative of the physical nature in which the sediment exists in suspension. For example, the LISST particle size measurements indicate bias towards the upper limits of the LISST measurement range. Generally, larger values within this upper region are associated with large particles outside the measurement range of the LISST (Agrawal & Pottsmith, 2000; Mikkelsen, 2002). Scattered light from these large particles leaks into the largest measurement cell, resulting in the observed bias. Video imagery recorded by divers during the frame deployment and retrieval confirm this hypothesis, with evidence of particle sizes exceeding 1 mm diameter in suspension. Particle size analysis of bed sediment surrounding the MI DPA and dredge material

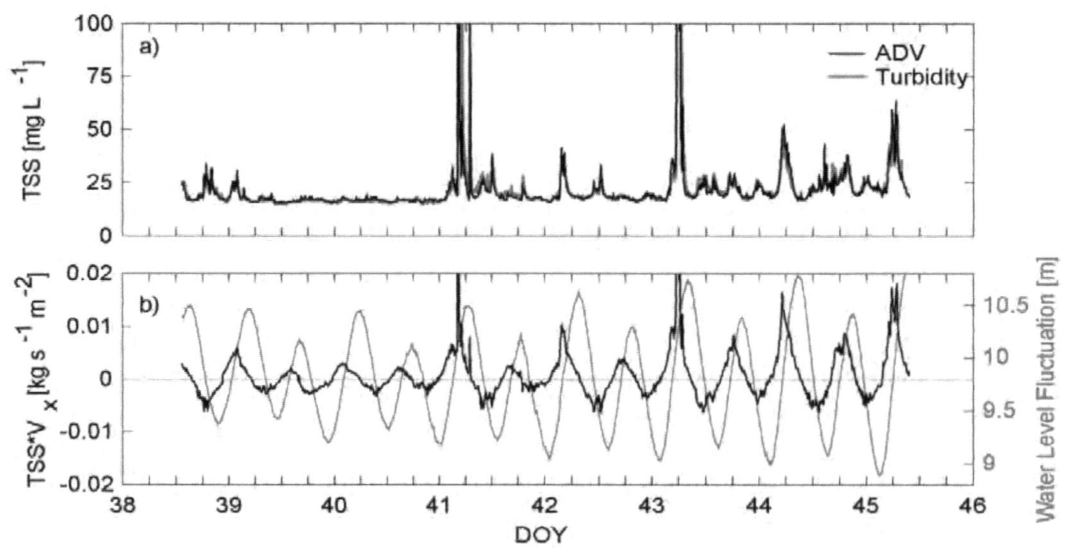

Figure 4: a) Calibrated TSS Concentration Time-Series b) Horizontal Concentration Flux Along the Principle Flood/Ebb Axis

placed within the DPA indicated that the material was comprised of mostly fine silts and clays with d_{50} in the range of 15 µm and 5 µm respectively (Beecroft et al., 2017; Lockington et al., 2016).

In alignment with the LISST measurements, TSS concentration estimates derived from the ADV BSI and turbidity sensor also identify a number of significant TSS peaks post day 41 (Figure 5a). Although the magnitude of these peaks are consistently smaller than those of the LISST, the locality and duration of such peaks align extremely well. Both the turbidity sensor and ADV were approximately 0.25 - 0.3 m above the LISST, hence TSS concentrations can be expected to be comparatively lower. One of the major advantages of the calibration of ADV BSI was the ability to measure TSS concentration fluxes at high frequencies 0.6 m above the bed. Figure 5b shows the TSS concentration flux per m^2 throughout the entire deployment period, with positive values referring to a flux along the flood tidal axis and negative values along the ebb tidal axis. TSS concentration flux results support the findings from the HR ADCP, with the largest flux estimates occurring during the flood tidal periods. The largest flux peaks occurred during days 41, 43 and 44, which also aligned with dredge placement events upstream of the ADV. This suggests that material already in suspension due to dredge placement operations or unconsolidated dredge material re-entrained from the bed may have contributed to the intensity of these larger TSS fluxes. As an estimate for net transport through the monitored site, TSS fluxes were integrated throughout the entire 7-day monitoring period, indicating a net transport of 610 tons per m^2 orientated along the flood tidal axis.

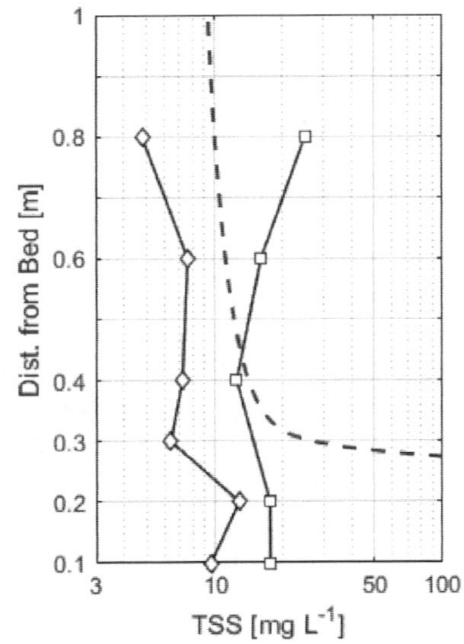

Figure 6: BBL Suction Profiler and Rouse Concentration Profile

Figure 6 presents the results from two BBL suction profiles taken during slack tidal periods on day 38 (squares) and day 45 (diamonds). TSS concentrations at 0.1 m above the bed in the range of 10 mgL^{-1} and 18 mgL^{-1} respectively. Characteristics of the profiles within this region indicated that little to no gradient in TSS concentration existed beyond 0.1 m above the bed. An alternate and frequently employed method to describe the vertical distribution of sediment refers to the simple Rouse formulation of concentration expressed as,

$$\frac{C}{C_a} = \left(\frac{H-y}{y} \cdot \frac{a}{H-a}\right)^{\frac{w_s}{\kappa u_*}} \tag{3}$$

Where C_a refers to the reference concentration at height a above the bed, H the water level, κ von Karman constant (=0.41) and C the concentration at height y *(Dey, 2014)*. Figure 6 presents this approximation (dashed line) throughout the neap tidal period spanning days 38 – 41 as a comparison to the BBL suction sampler results. The time averaged u_* estimates were derived from the HR ADCP profiles, while the reference TSS concentration and settling velocity estimates (w_s) were approximated from the LISST particle size distributions. The Rouse approximation provides similar estimates for TSS concentrations above 0.3 m from the bed, yet discrepancies between this theoretical approximation and our observations exists below this level.

4 DISCUSSION

Our observations demonstrate that the placement of fine silts and clays within the dredge placement area affected TSS concentrations and associated transport rates. Evidence for this is provided in Figures 4 and 5, where an increase in TSS concentration and frequency of concentration spikes were identified by all instruments at the onset of more intensive dredge placement operations after day 41. Generally, TSS concentration spikes were relatively short-lived and aligned with placement events located both upstream and within close proximity to the bottom-mounted frame. Although no direct measurements regarding the dispersion of dredge plumes during placement events is presented here, it is likely that the majority of deposited material is transported near the bed and remains in close proximity to the initial placement location (Gordon, 1974). Thus, observed short-term spikes in TSS concentration are believed to represent only a minimal percentage of the total quantity of dredge material placed within the MI DPA. Current estimates for the net transport of sediment through the site (610 tons per m^2) would be an important tool to confirm this hypothesis, yet information regarding the exact quantity (tonnage) of dredge material placed during the deployment period was unavailable.

Dredge material that does not initially settle to the bed remains in suspension as a turbid cloud. Particle sizes indicated that a large percentage of this turbid cloud was transported as larger flocculated particles. The direct correlation between TSS concentrations and d_{50}, combined with bias at the upper limit of the LISST measurement range were clear indicators for flocculation of the suspended sediment. Furthermore, video imagery (not shown here) also identified larger flocculated particles with particle diameters in the order of millimeters. Smith and Friedrichs (2011) presented similar results within dredge plumes, reporting increased particle diameters associated flocculation processes during dredge operations. As a result, settling velocities associated with these flocculated particles are expected to far exceed that of the primary particles (d_{50} = 5 µm), limiting transport potential of the suspended dredge material. While in-situ measurements regarding particle size and settling velocities are vital for the development of dredge plume and sediment transport models within coastal regions, evidence of these dynamic particle processes also gives indication to the location within the water column in which the majority of dredge material is transported. In the case of the MI DPA, previous concentration profiling periods confirms that the higher concentration suspension associated with the placement of dredge material was located in the lower 2-3 m of the water column (Beecroft et al., 2017).

The BBL suction profiler was developed as a tool to better understand the transport of material within the bottom-boundary layer, with previous monitoring attempts using alternate methods hindered due to extremely high TSS concentrations and an inability to place instruments close to the bed. Although the process is more tedious compared to optical and acoustic methods, suction profiling enables discrete TSS concentration estimates under extremely high concentration conditions and at a fraction of the cost. The results presented in Figure 6 provided valuable information regarding the distribution of suspended sediment within the near-bed region, indicating that theoretical approximations for the distribution of suspended sediment may not take into consideration the range of physical processes driving TSS concentrations. Hence, considerable differences were observed between TSS profiles obtained via suction sampling and those derived in accordance with *(3)* at distances less than 0.3 m above the bed. To improve understanding of transport rates within the MI DPA, measurements located closer to the bed (< 0.1 m), where TSS concentration fluxes are expected to be largest require further investigation.

5 CONCLUSIONS

Detailed measurements located within 1 m of the bed at the MI DPA unveiled key processes influencing the local transport of dredge material. Tidally driven currents dominate the net transport of suspended sediment, with a net flux orientated along the flood tidal axis. Furthermore, periods influenced by wind driven flows and an approaching spring tidal cycle were also attributed to an increase in transport potential from the site. The application of the LISST provided in-situ measurements related to the distribution of particle sizes, confirming flocculation as a major driver influencing the vertical distribution of suspended sediments. The ability to provide some fundamental understanding to the manner in which the dredge material is transported, is an important finding for the development of future sediment transport models in the area. Current modelling approaches which do not consider the gravity driven transport of material to the bed and the degree of flocculation would ultimately under predict the quantity of dredge material retained within the DPA.

This study has identified the need to extend both TSS concentration estimates and HR velocity profiling closer to the bed (< 0.1 m). Measurements within this region may be achieved through the application of high frequency acoustic devices such as ADVs or the HR ADCP. However, the BBL suction profiler is an extremely simple and cost effective tool for profiling TSS concentrations close to the bed, a method deployable under higher TSS concentrations compared to both optical and acoustic methods. More TSS suction samples are required under a broader range of hydrodynamic conditions and influence of dredge placement activities in order to provide representative estimates of TSS concentration fluxes at the site. Thus, future application of the near-bed monitoring system within the MI DPA and surrounds will form a key component of the ongoing research within Moreton Bay.

6 REFERENCES

Agrawal, Y. C., & Pottsmith, H. C. (2000). Instruments for particle size and settling velocity observations in sediment transport. *Marine Geology, 168*(1), 89-114.

Agrawal, Y. C., & Traykovski, P. (2001). Particles in the bottom boundary layer: Concentration and size dynamics through events. *Journal of Geophysical Research: Oceans, 106*(C5), 9533-9542.

Agrawal, Y. C., Whitmire, A., Mikkelsen, O. A., & Pottsmith, H. C. (2008). Light scattering by random shaped particles and consequences on measuring suspended sediments by laser diffraction. *Journal of Geophysical Research: Oceans, 113*(C4), n/a-n/a.

Andrews, S., Nover, D., & Schladow, S. G. (2010). Using laser diffraction data to obtain accurate particle size distributions: the role of particle composition. *Limnology and Oceanography: Methods, 8*(10), 507-526.

Beecroft, R., Cossu, R., Grinham, A., Perez, L., & Albert, S. (2017). *Cohesive suspended sediment observations in a dredge disposal area within Moreton Bay, Queensland.* Paper presented at the Australasian Coasts and Ports Conference, Cairns.

Cheng, R. T., Ling, C.-H., Gartner, J. W., & Wang, P. F. (1999). Estimates of bottom roughness length and bottom shear stress in South San Francisco Bay, California. *Journal of Geophysical Research: Oceans, 104*(C4), 7715-7728.

Dey, S. (2014). Suspended-Load Transport *Fluvial Hydrodynamics: Hydrodynamic and Sediment Transport Phenomena* (pp. 327-415). Berlin, Heidelberg: Springer Berlin Heidelberg.

Fettweis, M., Baeye, M., Francken, F., Lauwaert, B., Van Den Eynde, D., Van Lancker, V., . . . Michielsen, T. (2011). Monitoring the effects of disposal of fine sediments from maintenance dredging on suspended particulate matter concentration in the Belgian nearshore area (southern North Sea). *Marine Pollution Bulletin, 62*(2), 258-269.

Gordon, R. B. (1974). Dispersion of Dredge Spoil Dumped in Near-shore Waters. *Estuarine and Coastal Marine Science, 2*, 349-358.

Johnson, B. H., & Fong, M. T. (1995). Development and Verification of Numerical Models for Predicting the Initial Fate of Dredged Material Disposed in Open Water. Report 2. Theoretical Developments and Verification Results. In M. S. Army Engineer Waterways Experiment Station Vicksburg (Ed.).

Lockington, J. R., Albert, S., Fisher, P. L., Gibbes, B. R., Maxwell, P. S., & Grinham, A. R. (2016). Dramatic increase in mud distribution across a large sub-tropical embayment, Moreton Bay, Australia. *Marine Pollution Bulletin.*

Lohrmann, A., & Nylund, S. (2008, 17-19 March 2008). *Pure coherent Doppler systems - how far can we push it?* Paper presented at the 2008 IEEE/OES 9th Working Conference on Current Measurement Technology.

Mikkelsen, O. A. (2002). Examples of spatial and temporal variations of some fine-grained suspended particle characteristics in two Danish coastal water bodies. *Oceanologica Acta, 25*(1), 39-49.

Mikkelsen, O. A., & Pejrup, M. (2000). In situ particle size spectra and density of particle aggregates in a dredging plume. *Marine Geology, 170*(3), 443-459.

Nikora, V. I., & Goring, D. G. (2002). Fluctuations of Suspended Sediment Concentration and Turbulent Sediment Fluxes in an Open-Channel Flow. *Journal of Hydraulic Engineering, 128*(2), 214-224.

Onuf, C. P. (1994). Seagrasses, Dredging and Light in Laguna Madre, Texas, U.S.A. *Estuarine, Coastal and Shelf Science, 39*(1), 75-91.

Orpin, A. R., Ridd, P. V., Thomas, S., Anthony, K. R. N., Marshall, P., & Oliver, J. (2004). Natural turbidity variability and weather forecasts in risk management of anthropogenic sediment discharge near sensitive environments. *Marine Pollution Bulletin, 49*(7), 602-612.

Smith, G. G., Weitz, N., Soltau, C., Viljoen, A., Luger, S., & Maartens, L. (2008). Fate of fine sediment from dredger- based mining in a wave- dominated environment at Chameis Bay, Namibia.(Report). *Journal of Coastal Research, 24*(1), 232.

Smith, S. J., & Friedrichs, C. T. (2011). Size and settling velocities of cohesive flocs and suspended sediment aggregates in a trailing suction hopper dredge plume. *Continental Shelf Research, 31*(10), S50-S63.

ENGINEERING AND ENVIRONMENTAL IMPACT OF DREDGING FOR CONTAINER TERMINAL, BAY PORT, HAIFA, ISRAEL

[1]Noa Oren, Israel Ports Development and Assets Company Ltd. Tel Aviv, Israel
[2]Daniela Ostrovsky, Israel Ports Development and Assets Company Ltd. Tel Aviv, Israel
[3]Dan F. Di Castro, Civil & Marine Engineering Consultant, Haifa, Israel

ABSTRACT

The Israel Ports Plan includes the development of two focal points of sea trade along the Mediterranean Sea: at the city of Haifa located about 90 km northern to Tel Aviv and at the city of Ashdod located about 40 km southern to Tel Aviv (Figure 1).

The development plan was based on the forecast of cargos demand growth (import/export) and the raise of the global trade. One of the plan purposes was to enable huge vessels enter the Israeli ports. Logistics areas, transportation roads and railways were also taken into account.

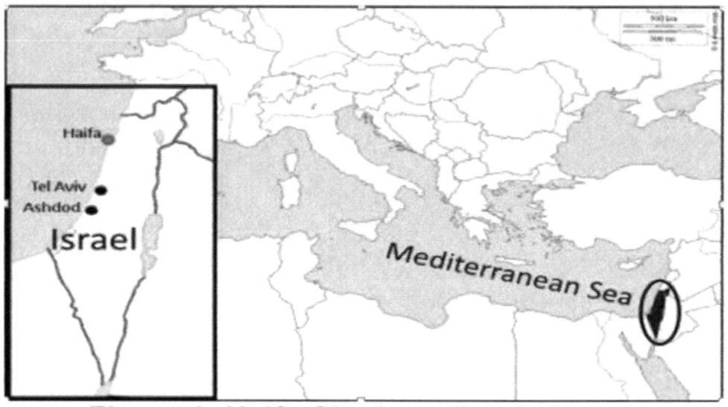

Figure 1: Haifa City Location in Israel

Bay Port, the new development stage, phase "A", is adjacent to the existing port of Haifa (Figure 2) built in the thirteen of the past century along the southern part of Haifa Bay. Further development stages of the port allowed berthing of 10 m draft cargo vessels in the sixteen, Panamax bulk carrier of 12 m draft in the eighteens and Post Panamax Container vessels of 14 m draft in 2007 including deepening of the entrance channel to -17.5 m.

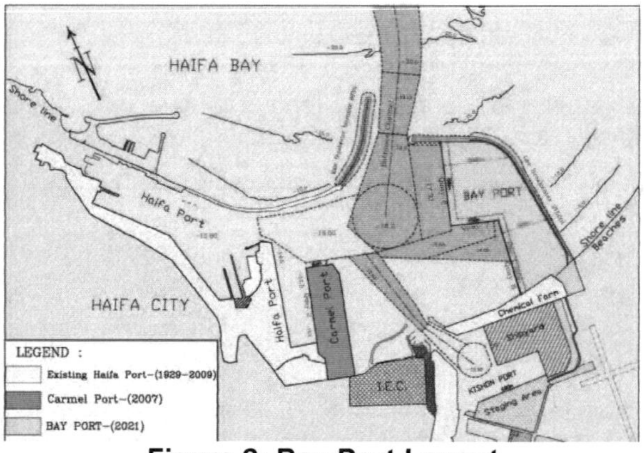

Figure 2: Bay Port Layout

Start order of Bay Port now under construction was on January 2015 and will be operational in 2021. Bay Port is designed for EEE container vessel (400x59x16m, 18,000 TEU) with an entrance channel of 20.5 m depth below Israel Land Survey Datum (ILSD). Total project cost is approximately 1 billion USD. The

[1]Noao@israports.co.il
[2]Danielao@israports.co.il
[3]famadica@netvision.net.il

project is executed concurrently with "Hadarom Port" project in Ashdod. Both projects are commissioned for Israel Ports Company Ltd (IPC), designed with the supreme supervision by HPA Engineers P.C.(USA) (ref.1). Bay port is under construction by the Contractor Ashtrom-Shapir Construction of a new Port Ltd. (Israel) and supervised on site by A.D.Y.R Constructions (Israel).
The location of the new port in a natural bay created engineering and ecological challenges during the design and construction phases especially in dredging and soil compaction aspects.
In this paper the following aspects which will be described are:

- Sand sea resources for fill reclamation,
- Deep compaction of the fill,
- Seismic impact and soil liquefaction,
- Impact of the new port on the adjacent beaches.

PROJECT DESCRIPTION

Bay Port includes the following main structures (Figure 3):

- 882 m extension of the existing Main Breakwater of the rubble mound type at water depth up to 16.5 m.
- 2,150 m of Lee Breakwater (1550 m of rubble mound and 600 m of rectangular cellular concrete caissons) at water depth up to -15m.
- 805 m main container Quay 6 for EEE vessels with design water depth of 17.3m. (Figure 4).
- 447 m secondary container Quay 7 for Post Panamax vessels with design water depth of 15.5m. (Figure 5).
- 715 m secondary Quay 8, of which 420 m for Panamax vessels with design water depth of 14.0m and 310 m with design water depth of 10.0 m for smaller cargo ships.

All depths are related to Israel Land Survey Datum (ILSD).

- 15,0 million m³ dredging(10.5 million m³ sand for the reclamation area of 840,000 m² , 3.1 million m³ of unsuitable material for fill (clay and silts) to be dumped offshore in Haifa Bay.
- For breakwaters, a total of 43,200 armor units of concrete Antifer cubes of 4,7 and12 m³ each and 5,500,000 ton rock of different sizes(0-1 ton,1-3 ton,3-6 ton)are required.
- A staging harbor 650 m long, with an area of 210,000 m² and water depth of 11.0 m and staging areas were allocated for contractor`s work.

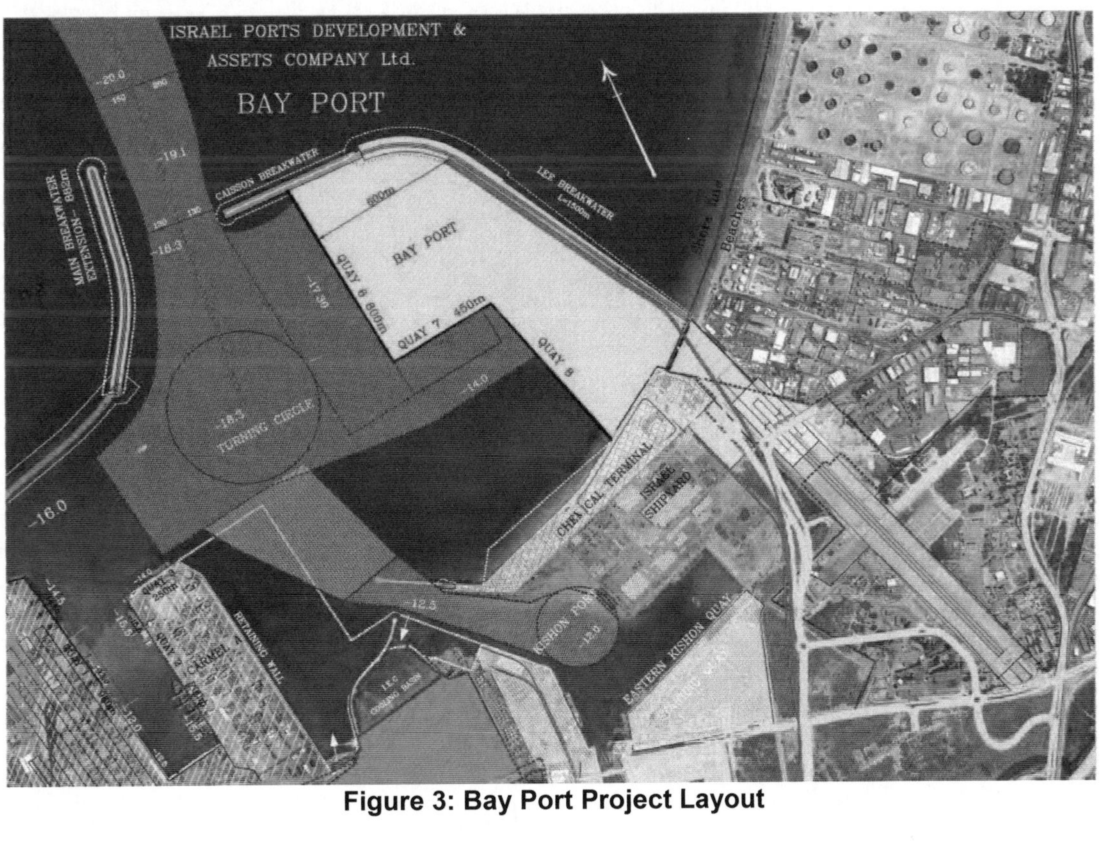

Figure 3: Bay Port Project Layout

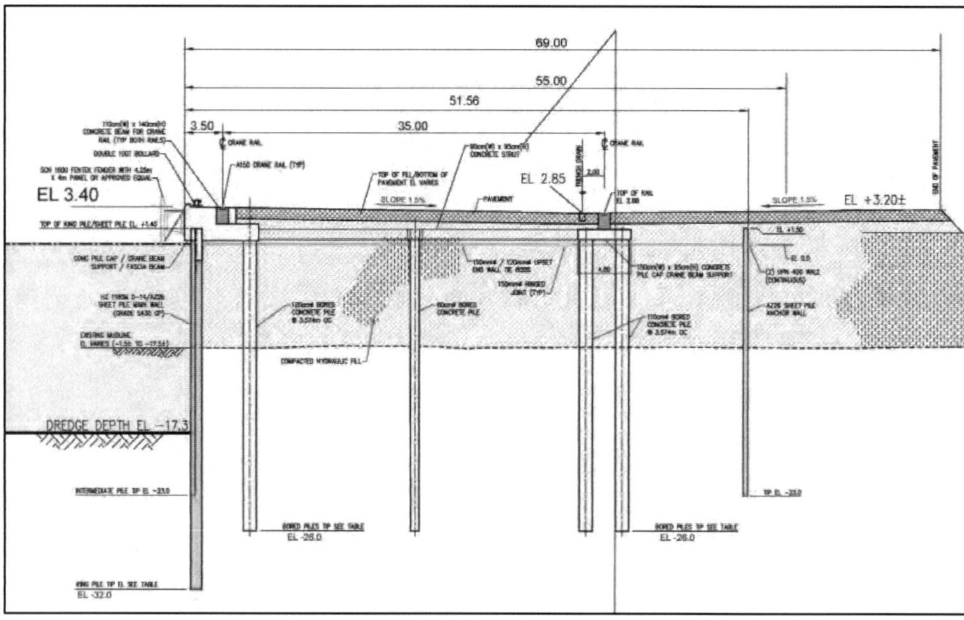

Figure 4: Quay 6 - Typical cross section

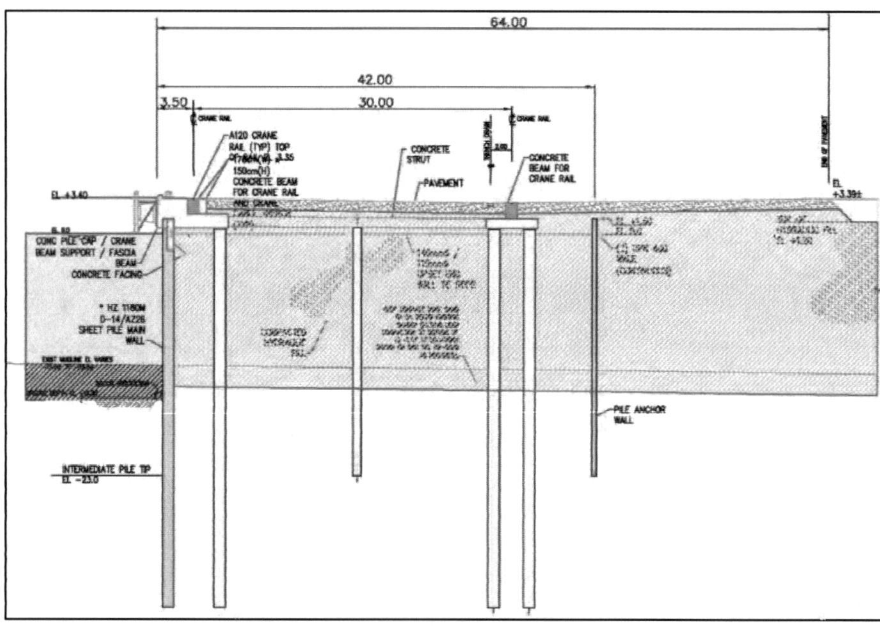

Figure 5: Quay 7 - Typical cross section

DESIGN

During the design phase of port's structures, extensive marine and land surveys were carried out. These mainly included: 135 exploratory boreholes, 66 CPT and 20 Vibro cores for soil investigations to depths up to -180 m below sea bottom; seismic and magnetometer surveys; soil laboratory tests; 2D and 3D hydraulic physical model tests of breakwaters stability; harbor wave agitation and ship motion numerical models; wave patterns at different construction progress scenarios; fast and real time maneuvering simulations for nautical design.

Four design features required a special study and attention for the proper and economic design of the project:

1) sand sea resources for fill reclamation,
2) deep compaction of the sand fill and settlement,
3) seismic impact and soil liquefaction,
4) short and long-term impact on the adjacent beaches of Haifa Bay due to the Bay Port new breakwaters,

entrance channel and the area of sand mining close to the project.
This paper shall focus on the above issues.

DESIGN FEATURES

1. Sand resources for fill reclamation

The extension of Bay Port involves the construction of new container stacking yard, mostly reclaimed over the sea by walls and retaining structures.
For the reclamation area, huge quantities of fill material were required. An economic solution is to use fill material to be found as much as possible in an area close to the reclamation site.
The quantity of the sand fill material to be reclaimed inside the perimeter of the sheet piles walls is ca 10.50 million m³. This quantity includes 3.65 million m³ of sand to be dredged in an area delimited from the deepening to the operational depth in the entrance channel and in the approach channel, turning circle, along quays 6, 7 and 8.The remaining quantity of 6.85 million m³ of sand has to be dredged from a borrow area outside the port in Haifa bay. The suitability of the sand was determined on the basis of the following criteria:

- The sand fill must be able to withstand the designed loads.
- Start of operations immediately on completion of the fill and other works.
- Supply of the sand according to the planned schedule and fill procedures within a period of two years.
- Environmental considerations.
- Minimum price
- Statutory and legal aspects.

1.1 Following the National Master Plan for Marine Structures Works instructions, IPC prepared an Environmental Impact Assessment (EIA) including an Environmental Monitoring and Management Plan (EMMP) for dredging works. The EIA was approved by the Ministry of Environmental Protection (MOEP).
Extensive studies were performed by international experts (ref.2) as well as Israeli specialists (ref.3) in the fields of hydrodynamics, sedimentology, biology and chemistry. The EIA is implemented during the construction.
The (EIA) of Shallow Water Dredging has included the following main issues:

- Baseline Investigations and Surveys;
- Modelling in the field of waves, hydrodynamics and sediment transport (2D), sediment spill and water quality;
- Short-Term Impact of dredging at alternatives sites using various technologies;
- Long Term Impact on the marine environment after completion of the dredging work;
- Selection of the preferred alternative;
- Monitoring, Management and Mitigation plan to ensure compliance with the environmental quality objectives defined for the project.

Along the Israel coast sand deposits to be considered as potential borrow areas generally contain high fines content (i.e. the percentage of material passing 0,075 mm sieve)which is unsuitable for engineering purposes as hydraulic fill for reclamation when low fines content is required .Two sand mining alternatives for the construction of Bay Port have been examined:

- Alternative 1- Sand mining solely from the Haifa site.
- Alternative 2- Sand mining from Haifa and Nitzanim site.

Dredging in each of the following sites was examined:

a) Haifa site is close to the reclamation area situated in the Bay where an unique ecological system exists (Figure 6), The granulometry of material to be dredged in this site mainly consist of very fine sand. The sand layer overlies grey dense clay in most of the survey area. The composition of the material is mostly quartz sand with some silt and local clay lenses. In addition to 9 boreholes and 18 vibrocores executed in the borrow area, an accurate Seismic survey was also executed in the same area to verify the sand thickness layer.
b) Nitzanim site lies south to Ashdod Port, distant from Haifa Bay Port about 140 km (Figure 7). The granulometry of material to be dredged in this site consist of fine to medium sand. To verify the sand thickness layer, 16 vibrocores were executed in this site.

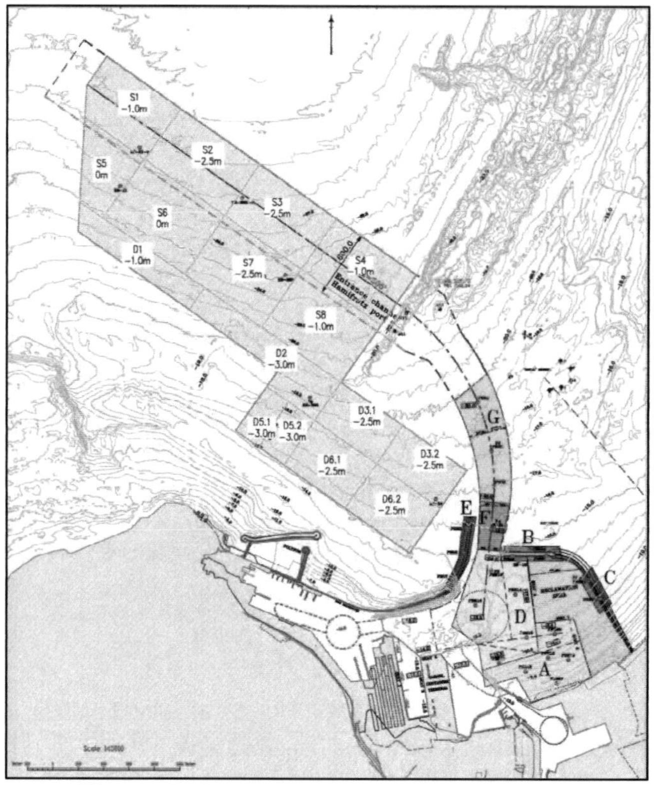

Figure 6: Borrow Area at Haifa Bay Site

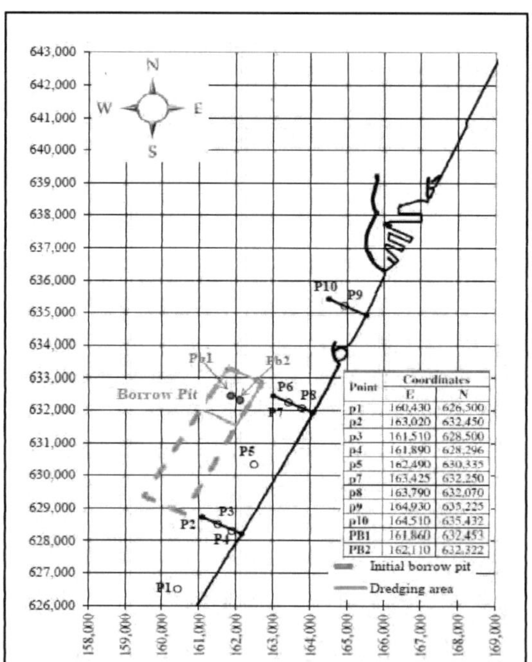

Figure 7: Borrow Area at Nitzanim Site

Some 1.25 million m³ of dredged material was estimated to be available from Nitzanim site and the quantities to be dredged from the Haifa borrow pit thus can be reduced accordingly.

The total quantities of fines (in tons) from spillage due to dredging and reclamation, computed from the two dredging sites, were compared for short-term as well as for long-term impacts .

After having examine technical, environmental and economic aspects of the two alternatives for sand dredging, Haifa site alternative was chosen taking into account its advantages and was implemented in

the construction works.
The sand from Haifa borrow area adjacent to the reclamation works was dredged at water depths of -16.0 to -18.0 m below the Israel Land Survey Datum (ILSD); the dredged layer thickness varied between 1.0 to 3.0 m (Figure 6). Dredging, for the whole project, was executed by TSHD equipped with green valve.

1.2 Water turbidity control
The execution of the Works has been carried out in such a way as to minimize adverse impacts on the existing sea water quality and marine ecology.
The dredging works specifically included partial loading (reduced capacity) of TSHD to limit the quantity of soil spilled during dredging and reclamation.
The Turbidity Monitoring includes:
• Satellite Images Monitoring: monthly at high resolution and daily at low resolution.
• Continuous On-Line Monitoring by Sensors during all the dredging and reclamation period. Eleven on line turbidity sensors were deployed as indicate in Figure 8 .The records are every ten minutes and transmitted to shore station.
• Daily Turbidity Monitoring by boat were performed at 7 points, at 2 depths at each station (1 m below surface, mid-depth). (Figure 9).

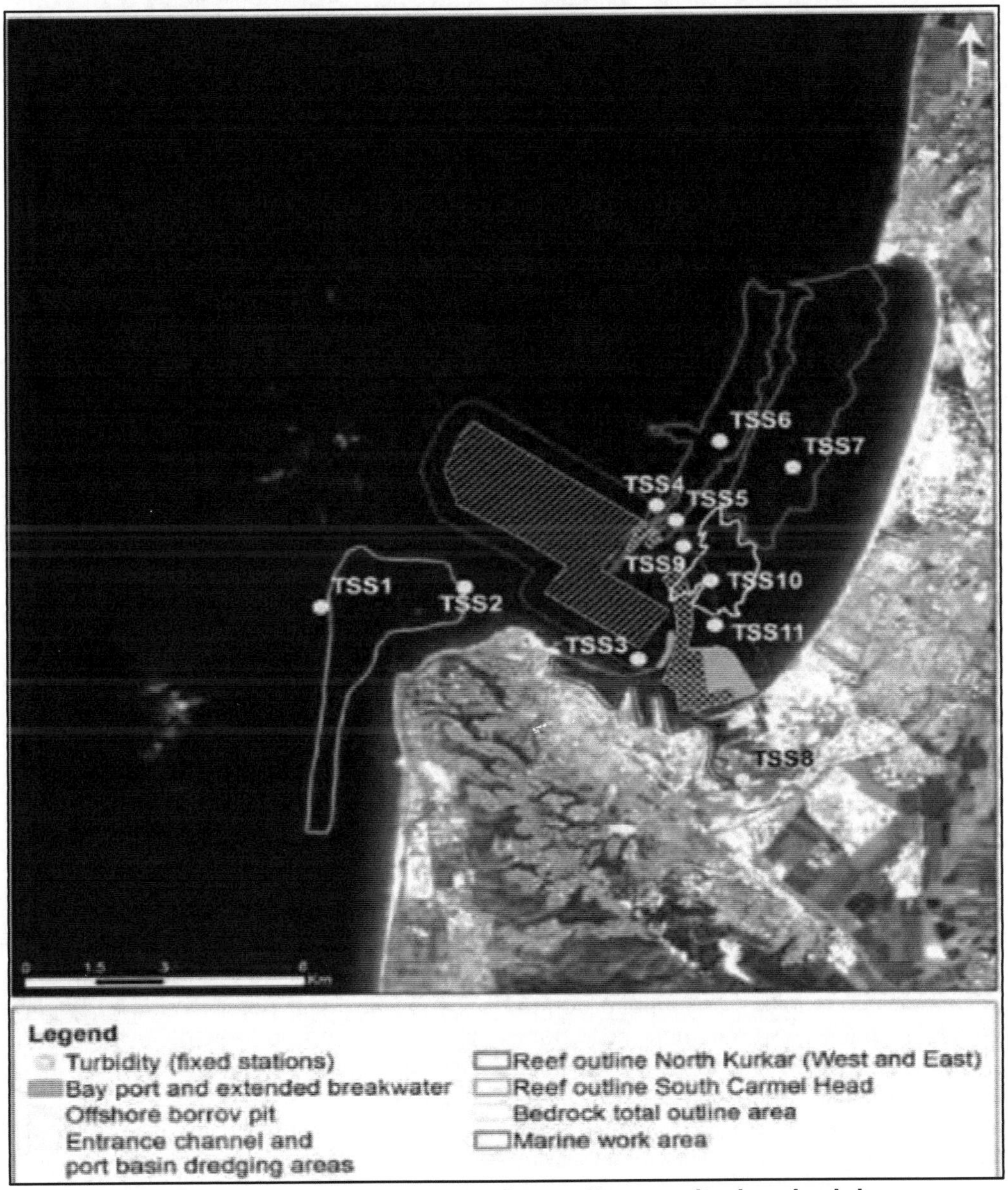

Figure 8: 11 turbidity sensors for on-line monitoring dredging

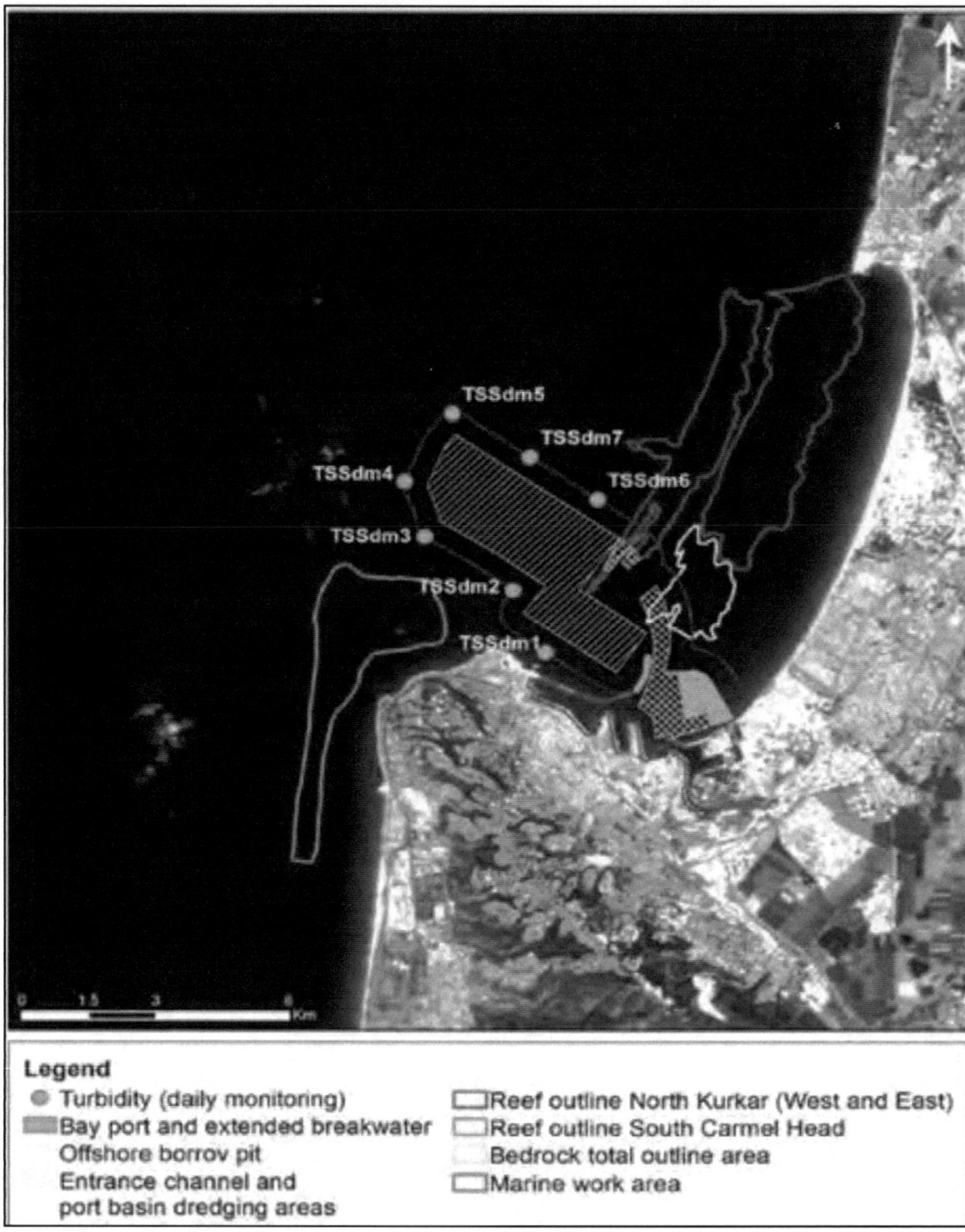

Figure 9: 7 Monitoring stations for daily turbidity monitoring by boat

The maximum sediment spillage fully compliant with the allowable spillage was defined in the EMMP as follows:
For stations TSS9, TSS10 and TSS11 (at half depth) continuously monitored: 30 mg/l.
For stations TSSSdm1 to TSSdm7 (1 m below surface and mid-depth) monitored by boat: 30 mg/l.

1.3 Procedure placement of suitable loads in the TSHD hopper.
According to the Specification of the Work, the average fines content in the reclamation fill should not exceed 23%. Suitable materials in the hopper were considered loads that have a fines content not exceeding 27%.Loads containing fines whose content exceeds 23% but not exceeds 27% are allowed to be placed selectively in the reclamation in order not to exceed an average fines of 23%.When the sample contains less than 23% fines the load will be placed in the reclamation according to the placement schedule. When the sample contains more than 27% fines, the load will be disposed offshore, in disposal area "Epsilon" approved by the MOEP for unsuitable-not contaminated materials. The distance of Epsilon site from shore is about 22.5 km. For contaminated materials, the disposal area is "Alpha" also approved by MOEP at a distance from shore of about 40 km. (Figure 10).

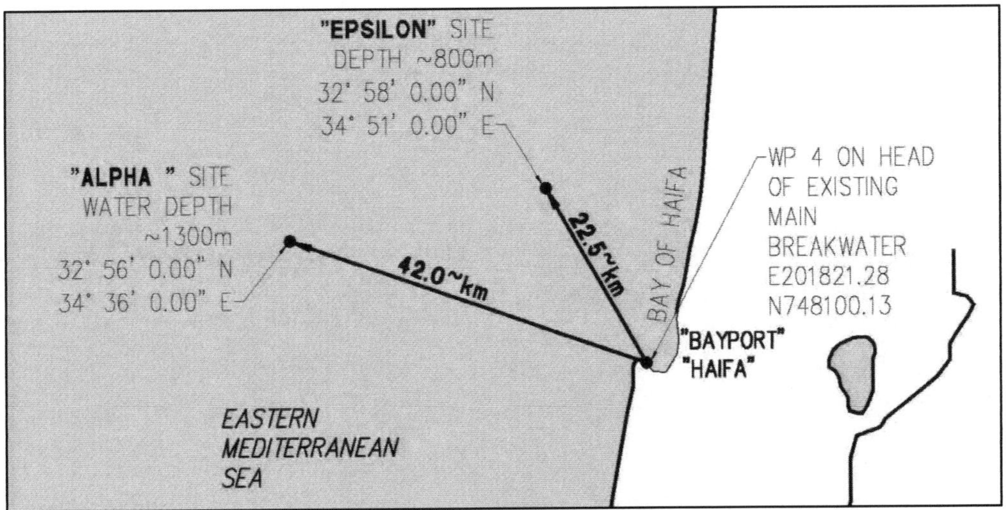

Figure 10: "Epsilon" and "Alpha" disposal sites opposite Haifa coast

2. Deep compaction of the sand fill in Reclamation Area – Settlement calculation and monitoring

The reclamation behind the new quay walls required the use of ground improvement works to accelerate and mitigate the ground movements associated with reclamation works, reduce future settlements and mitigate the risk of liquefaction in case of seismic event.

Two typical geotechnical sections of the various layers underneath the reclamation area are shown (Figures 11 and 12).

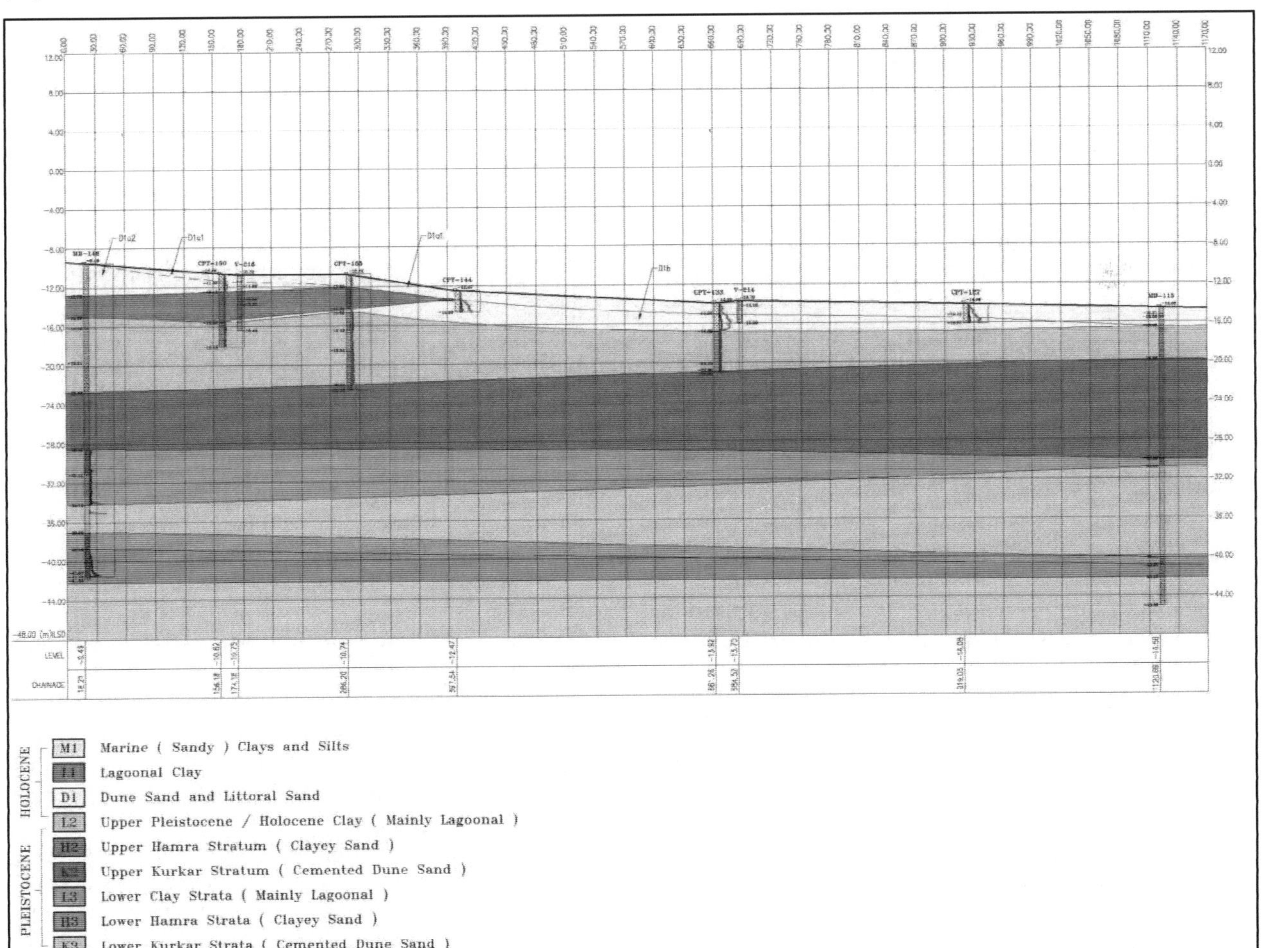

Figure 11: Geotechnical section along Quay 6

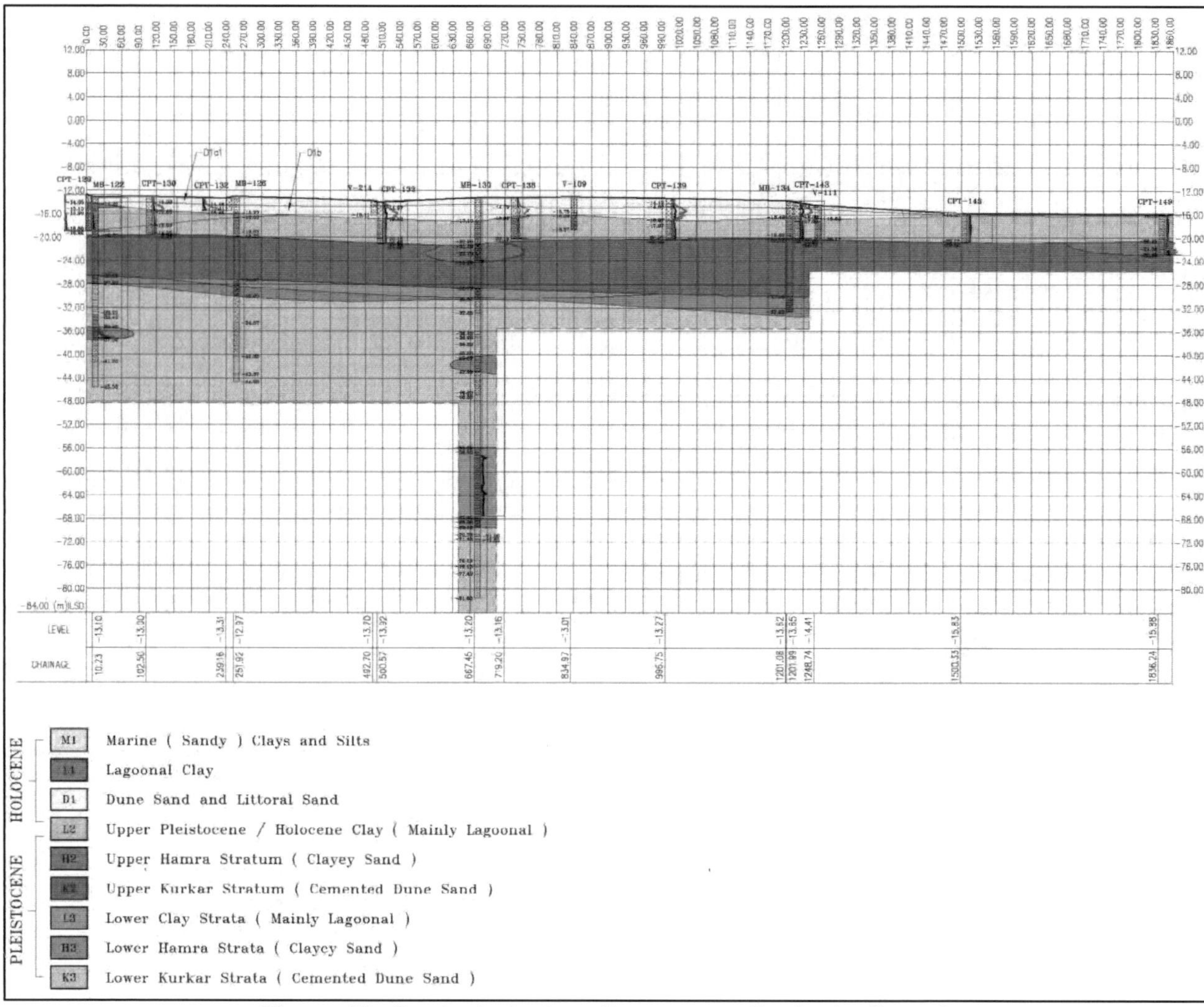

Figure 12: Geotechnical section perpendicular to Quay 6

Based on the results of soil investigation program, the stratigraphy of the project site is typically as follows: Dune sand, Upper Lagoonal Clay (L1), Upper Hamra (local term for clayey silica sand, cemented), Upper Kurkar (local term for uncemented sand to strong calcareous sandstone), Lower Lagoonal Clay (L2), Lower Hamra, Lower Kurkar.

Reclamation fill is expected to undergo elastic settlement as well consolidation settlement due, to the upper lagoonal clays L1 of very soft consistency and L2 stiff clay.

With the aim to minimize expected settlements due to primary and secondary consolidation of the sea bottom clay layers and liquefaction of the sand due to seismic effects, ground improvement techniques of deep compaction like Vibrocompaction and /or Vibrostone columns(using the wet top-feed process) were implemented from 2 m below the existing seabed up to +0.5 m ILSD..

Preparatory works like Boreholes and Proof Testing 30 by 30 m grid were performed for data analyses and determining compaction methodology.

If Fines Content (FC) in the fill area was found less than 23%, works were carried out according to Spec ification. If FC was found more than 23%, Proof Testing with grid 10X10 m were performed for data analyses and determining compaction method and grid .If (FC) was found more than 10% but less than 23%, or less than 10% and based on laboratory test results, (PSD ,carbonate content, pCPTs) choice was made between vibrocompaction and Vibrostone columns.

The criteria for compaction were divided into three main types (Figure 13) based on Area Replacement Ratio (ARR) which includes both static and seismic considerations. ARR is defined as the ratio of the stone columns area to the total area of the soil/stone matrix. The three types were:

Type 1 - along quays no.6, 7 and 8 and buildings foundations was compacted by Vibrostone columns with ARR of 23%.

Type 2 - the biggest internal area for containers storage, roads etc. was compacted by Vibrocompaction (70% relative density) and/or Vibrostone columns with ARR of 12%.
Type 3 - intended for dangerous and hazardous containers storage was compacted by Vibrostone columns with ARR of 25%.
The fill behind the quays and revetments was compacted to mitigate the effects of liquefaction under the 475 year return period earthquake and beneath the dangerous and hazardous container area was compacted to mitigate the effect of liquefaction under the (2/3) of 2,475 year return period earthquake.

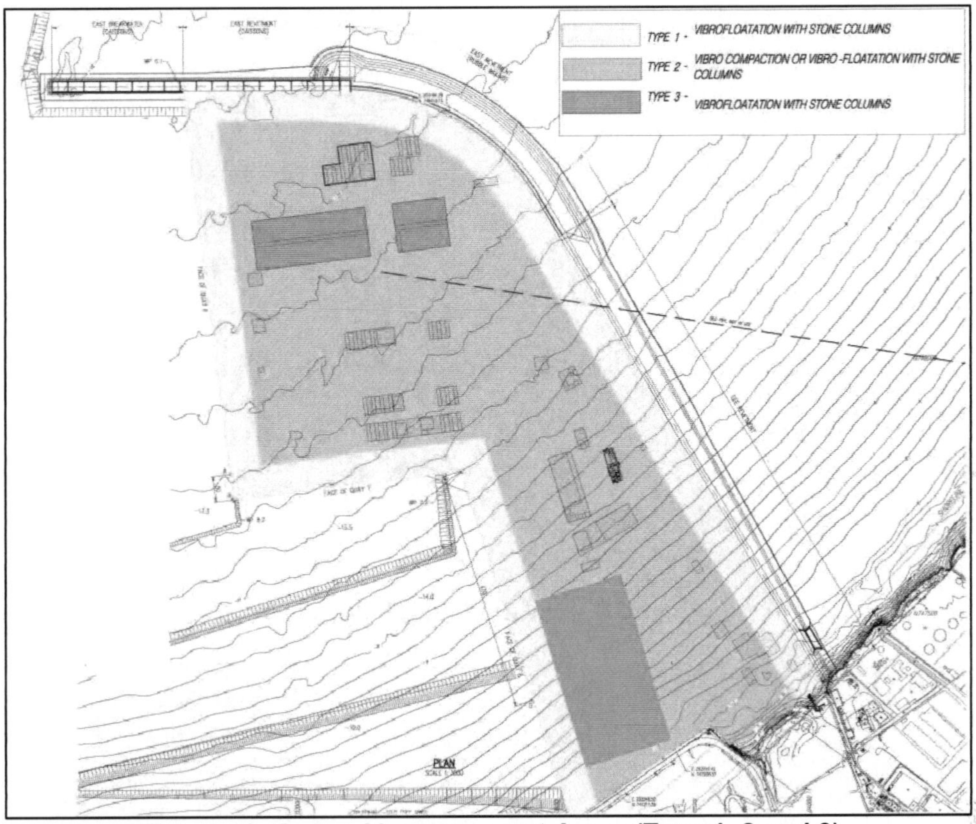

Figure 13: Ground improvement Areas (Type 1, 2 and 3)

Stone columns in area Type 1 were placed on a triangular grid with center to center of 2.5m or rectangular grid of 2.40 x 2.25m due to tie rods position between main and anchor walls and stone column dia. of 1.26m .In area Type 3 a triangular grid of 2.5m was adopted and stone column dia. of 1.32m executed. Vibrostone columns area Type 2 was carried out with a triangular grid, center to center 2.5m and dia. of 0.91m.
After a series of additional Proof Testing by means of pCPT, the values of 23% ARR in area Type 1 and 25% in area Type 3 were reduced to 20% ARR, provided the FC distribution within the reclamation shall be found within 10% to 12%. Stone column diam. was 1.17m.
Time dependent settlements were estimated at a total 35 boreholes and CPT locations spread over the reclamation area. These settlements arise as a result of consolidation as well as secondary compression. For the calculation of the ultimate settlement total overburden stress applied at the mud line and a permanent surcharge of 25 kPa were taken in account. It was assumed that construction time to reach the required fill height is about 1 year.
The potential remaining settlement 2 years after reclamation fill placement due to upper lagoonal clay. This 2-year post-construction settlement is largely confined to the south portion of the reclamation area. Elsewhere, barring a few locations at the East Lee Breakwater, the settlement is practically nil. Some of the total 31 settlement plates for settlement monitoring have been already installed in the reclamation area. Surveys of elevations of the settlement plates have been carried out on a monthly basis including cumulative settlements.
Up to date positive and acceptable results were achieved by both compaction methods, Vibrocompaction or Vibrostone columns.

3. Seismic impact and soil liquefaction

3.1 The northwestern coast of Israel lies in a seismically active eastern Mediterranean region and Haifa Bay is the most active zone along the Israel coast .In the past several large earthquakes have occurred up to a magnitude of 5.3. The main reasons for seismic risks are the proximity of the Carmel (Yagur) fault. Its location within the Kishon graben is shown (Figure 14).

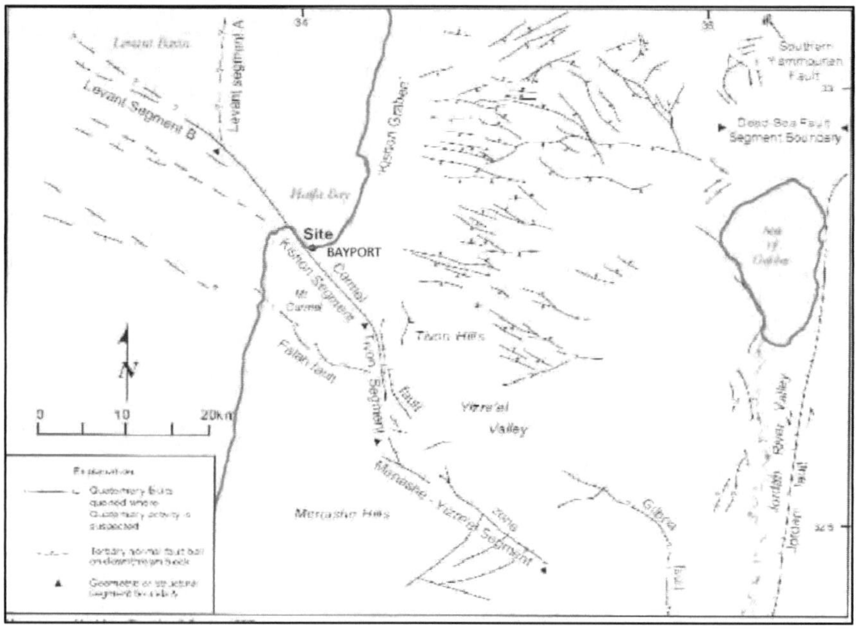

Figure14: Carmel fault location (ref.4)

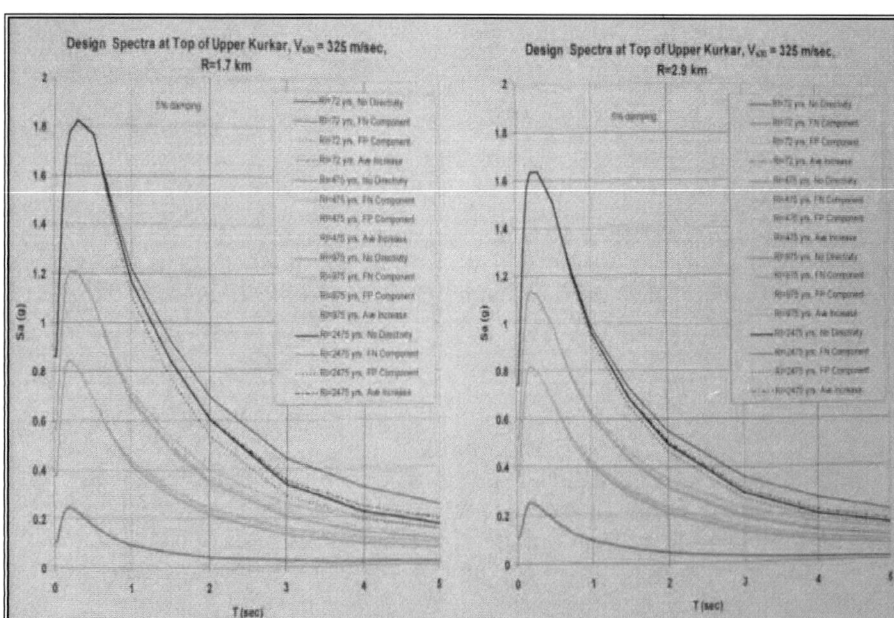

Figure15: Design spectra for OLE and CLE

The primary goal of the ground treatment due to an earthquake is to minimize ground deformations to acceptable levels near to critical structures like breakwaters, quays, ship to shore cranes, utilities systems, hazardous cargo storage facilities, upland utility corridors etc.

For the development of Bay Port, (IPC) decided to implement ground treatment over the whole terminal area. The extent and degree of ground improvement were planned in such a way to reflect the importance of the specific location and resources to repair the facility following seismic events.

Design response spectra were developed by the Designer (ref.1 and ref.4) based on seism-tectonic studies and ground shaking analyses and time histories developed for two design levels of return probability, the Operating level earthquake (OLE) with a return period of 1 in 72 years, 50% probability of exceedance in 50 years and a Contingency level earthquake (CLE) with a return period of 1 in 475 years,10% probability of exceedance in 50 years .For the OLE event, a PGA near the top of the upper rock stratum of 0.11g and for the CLE event a PGA of 0.38g were used for the design of Bay Port structures.
Hazardous and Dangerous Container area was located sufficiently upland of the quay face so that this area will not be subjected to a quay related damage. The adopted PGA was 0.57g. Plots of the design spectra for seismic events are shown in Figure 15.

3.2 Soil Liquefaction and its control was a major design consideration in Bay Port project due to the use of hydraulic sand fill for reclamation and the relatively severe design earthquake conditions at Haifa Bay .A relatively density of at least 70% was assumed to be required to prevent liquefaction .Sand placed underwater has very low density and therefore generally is subjected to liquefaction.
In order to eliminate the potential for liquefaction, the soil at any given depth must have a Cyclic Resistance Ratio (CRR) greater than Cyclic Stress Ratio (CSR) at the same depth. A safety factor (SF liq) of 1.15 has been adopted in developing the required CRR.
As part of the Liquefaction Assessment as required by the MOEP, cyclic loading due to CLE quake motion and cyclic resistance of reclamation sand fill were tested in laboratory and analyzed. The Liquefaction Assessment used standard empirical methods and assumed soil parameters for the hydraulic fill and natural sand layers. The conclusion was that liquefaction will occur under the design earthquake conditions: a) in the sand fill reclamation: and b) along parts of the foundation of the main and lee breakwater unless appropriate mitigation measures should be implemented.
For sand fill reclamation, deep compaction techniques were used as described in design feature 2 (Deep compaction of the sand fill in Reclamation Area – Settlement calculation and monitoring). For foundation of main and lee breakwaters, replacement of liquefiable upper Dune sand layer with stone of "E" class(0-1000kg) was executed along the main breakwater extension and partially, where considered necessary, along the lee breakwater. (Figures 16 and 17).

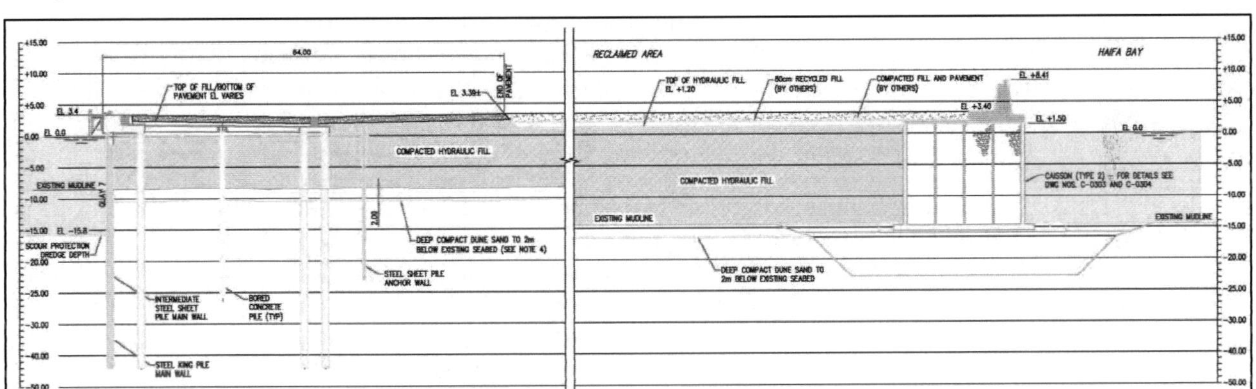

Figure 16: Typical section of reclamation area parallel to Quay 6

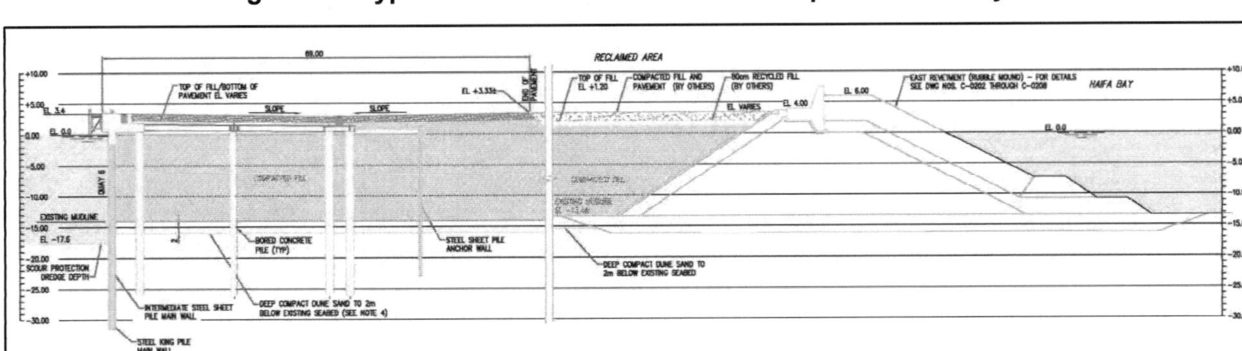

Figure 17: Typical section of reclamation area perpendicular to Quay 6

4. Short and long term impact on the adjacent beaches of Haifa bay

4.1. Haifa Bay can be considered as the end of the Nile littoral cell. The input of Nilotic quartz sand into Haifa Bay is therefore smaller in quantity in comparison with other parts of the Israeli coast, In Haifa Bay due to its location in the sheltered area of the Carmel headland the wave energy is reduced by a factor of 3-4 compared to the Haifa southern coast. The annual sand transport which enter the Haifa Bay from the south coast is estimated at 80,000 m³. For the present layout, the amount of sediments that bypasses the Main Breakwater and possibly enter into the littoral zone of Haifa Bay is ca. 3900 m³/year. Stirring and movement of sediment in the Haifa region is primarily generated by longshore currents induced by breaking waves as well as turbulent velocities generated by the breaking. The direction of sediment drift depends on the angle of the incoming waves measured relative to the coast normal.

The Environmental Impact Assessment (EIA) submitted by the IPC to the Ministry of Environmental Protection (MOEP) dealt with the impacts in Haifa Bay due to the extension of main breakwater, building of the lee breakwater , dredging at the entrance and approach navigation channel and sand mining in the borrow site to be used for fill in the reclamation area.

The study of those impacts commissioned by IPC to CAMERI (ref.3) included:

- comparison of bathymetric and topographic maps and ortho-photos,
- comparison of cross shore profiles,
- changes in wave climate conditions due to the construction of the new structures,
- location of strong currents,
- changes in sediment transport,
- coastline evolution,
- erosion close to marine outlets,

The most advanced and update models for flow and morph dynamic computations were used by CAMERI (ref.3).

The study considered two layouts of the Haifa Port- existing port used, as reference state, and designed port–Bay Port.

The area of interest in terms of sediment transport modeling stretches along the shore from the northern port of Atlit ruins in the south to Old Acre town in the North (a length of ca.25 km) with the western boundary limited by the 30 m water depth (Figure 18).

The wave climate conditions utilized are based on wave instrumental measurements by Datawell wave rider positioned at Haifa site, at water depth of 24 m. The wave dataset covers 16 complete hydrographic years. Based on the analyses of wave data, ten representative waves were selected to represent adequately sediment transport potential within the area of interest (Figure 19) and cross profiles (F1 to F9) drawn in predetermined locations and compared among them (Figures 20 and 21).

The coastline evolution has been assessed during a 30-year period after the Haifa Bay Port construction (Figure 22). Results of the coastline evolution were also assessed every 5 years. The extension of the main breakwater by 882 m was expected to reduce the bypass rates already small at present and consequently to increase beach erosion along the bay since the north-going transport would no longer be compensated, even partially. In such conditions sand nourishment, of about 35,000 m³/yr, was anticipated based on models results.

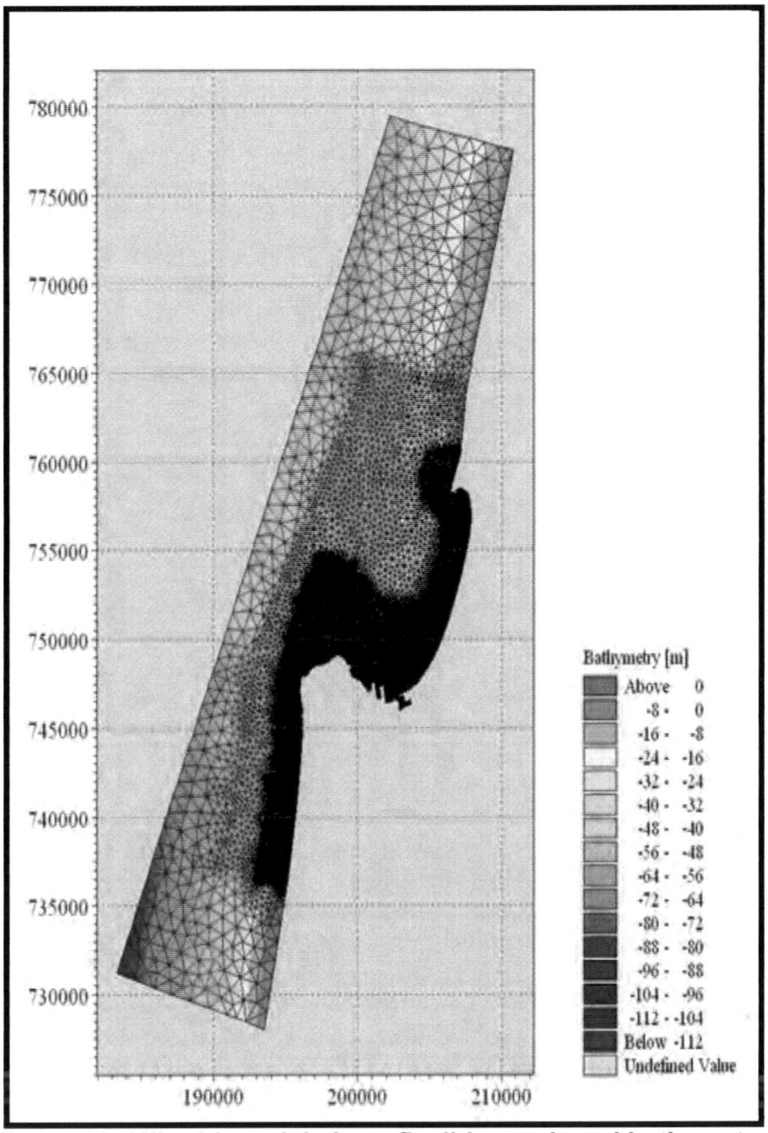

Figure 18: Mike 21 model shore flexible mesh and bathymetry

No.	Waves			Wind	
	Significant height, H_{m0}	Direction, *Dir*	Peak period, T_p	Speed	Direction, *Dir*
	m	°N	sec	m/s	°N
W01	2.5	245	7.7	7.5	240
W02	2.6	252	6.7	8.2	250
W03	2.8	259	11.1	8.5	259
W04	3.1	264	10.0	8.0	270
W05	3.6	268	9.1	9.0	268
W06	4.7	272	11.8	9.5	274
W07	3.2	285	11.1	6.0	220
W08	4.2	291	11.8	11.0	260
W09	2.9	299	9.1	8.9	299
W10	2.6	310	8.3	10.0	330

Figure 19: Representative waves parameters in deep water

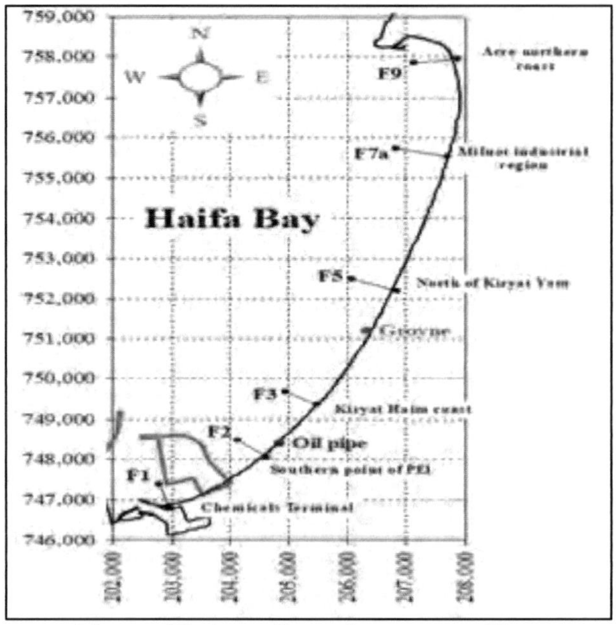

Figure 20: Location of cross profiles

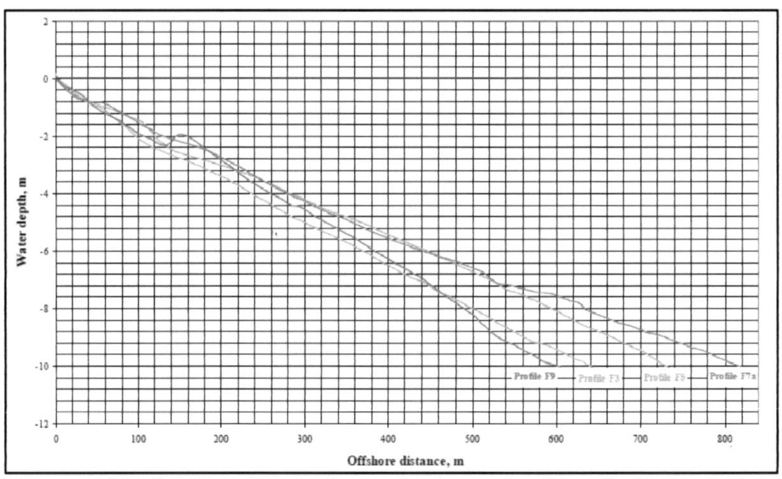

Figure 21: Comparison of cross shore profiles in Haifa Bay

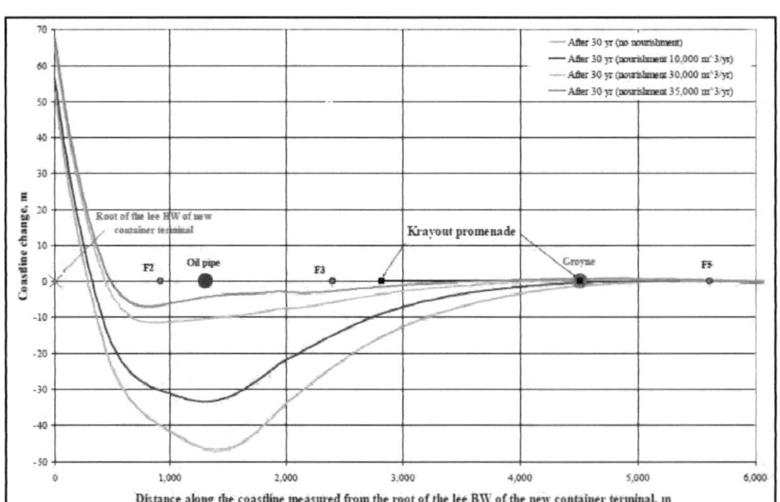

*Negative values refer to coastline retreat, positive refer to coast line advancing.

Figure 22: Coast line alignment 30 years after the Bay Port completion

4.2 The sedimentological issue and impacts were part of the short and long-term Sedimentological Monitoring Program.

The Monitoring phase of sand nourishment yet started along the Haifa Bay beaches during the construction phases due to some erosion along the beaches northward to the lee breakwater, especially in front of rigid structures close to the shore line like long concrete fences, promenades, ramps, drainage outlets etc. Result of natural sand accretion northward to the Lee breakwater is shown in Figures 23 and 24. Artificial sand nourishment was carried out by TSHD pumping sand on the shore by means of floating pipe from the dredger and diffuser moved along the beach, following a predetermined nourishment plan.

Borrow areas of sand of similar quality and gradation to that of Haifa Bay beaches were found opposite the old main breakwater in water depths of 10 to 16 m (Figure 26).

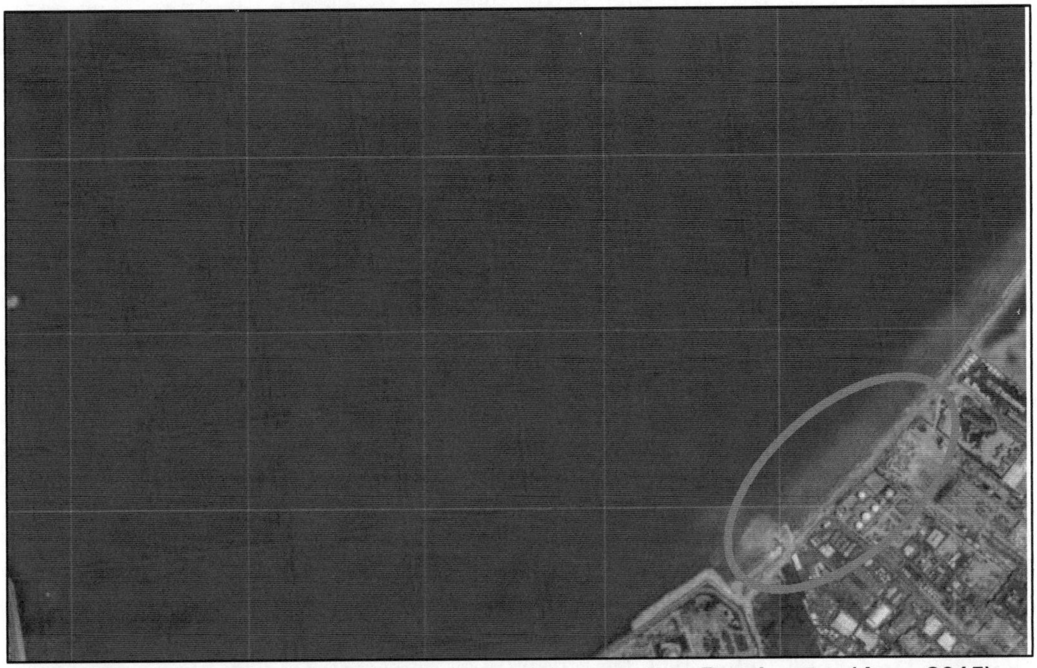

Figure 23: Starting the construction of the new Lee Breakwater (June 2015)

Figure 24: Beach accretion northward to the Lee Breakwater (February 2018)

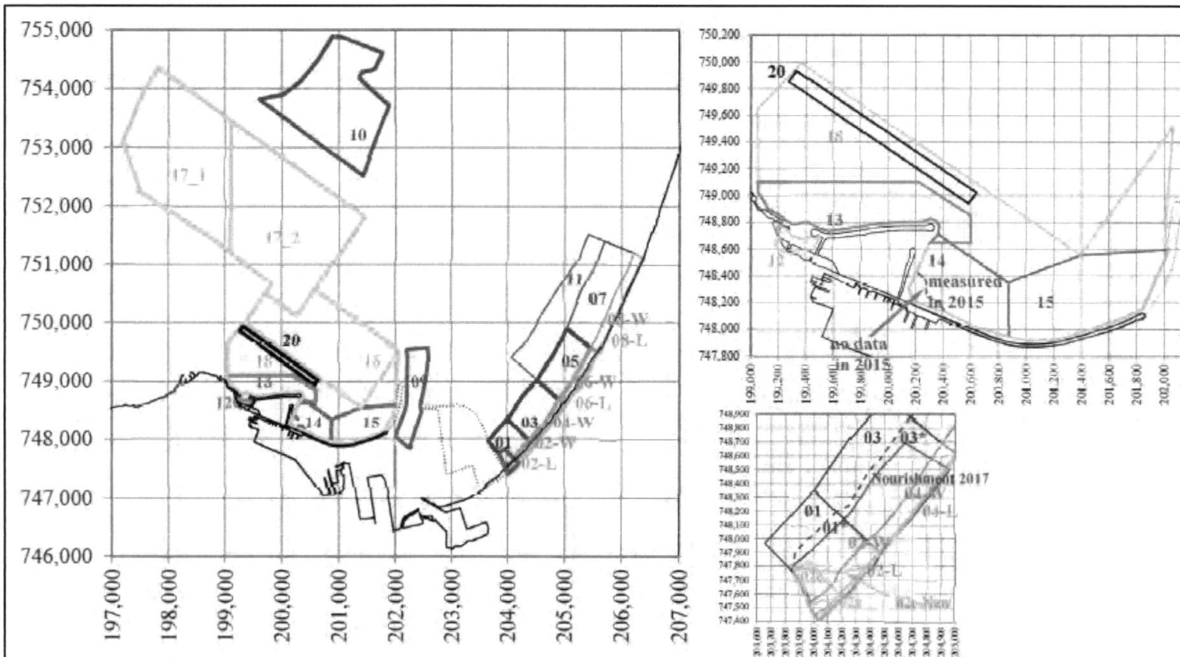

Fig 3.1.1. – Location of the considered lots.

Figure 25: Haifa Bay and Bay Port - Location of the considered lots

Analyses of bathymetric maps considered the changes occurred during summer and winter of years 2016, 2017 and partially 2018 and comparison of predetermined lots (i.e those lots that cover the borrow area and the dumping area). Volumes, active area and average depth changes in each lot were computed and analyzed thus obtaining sand accretion or erosion in each lot. Topographic and bathymetric surveys were executed along the coast using single-beam and multi-beam echo sounder.

The following sand nourishments campaigns were carried out, close to the shore line, until the end of 2017 in the following lots (Figure 25):

April-May 2016 70,870 m³ along lots 06-08
September-October 2016 100,050 m³ along lot 06
May-June 2017 179,110 m³ along lots 05-06
September-October 2017. 7,600 m³ along lots 05-06

Results obtained through comparison and analyses of differential maps indicate on sand accretion, as expected, close to the northern part of the Lee Breakwater under construction and some seasonal beach width improvement along the remainder shore where sand nourishment was carried out opposite the eroded sectors.(Figures 27 and 28).

Figure 26: Trailer suction hopper dredger pumping sand ashore

Figure 27: Beach width In front of concrete fence before sand nourishment

Figure 28: Beach width In front of concrete fence after sand nourishment

CONCLUSIONS

The construction of Bay Port is not yet fully completed, but the following preliminary conclusions can be reached:

1. The sand from Haifa borrow area adjacent to the reclamation works was dredged in such a way to minimize the existing sea water quality and marine ecology following a predetermined strictly monitoring.

2. Positive results were achieved for reclamation works by both compaction methods i.e Vibrocompaction or Vibrostone columns although the fines content in the sand to be dredged for the reclamation fill sometimes reached 25%.

3. The foundation of Main Breakwater and Lee Breakwater (rubble mound) were treated by removing the liquefiable Dune sand layer and in such a way minimizing soil liquefaction during earthquake.

4. During the construction phases, sand nourishment along the Haifa Bay beaches lied northward to the Project together with continuous monitoring carried out by hydrographic, topographic mapping and aerial photos improved the width of limited previewed eroded sectors along the sea shore northward to the Project.

REFERENCES

1. HPA Engineers P.C. N.Y, USA. New Marine Container Terminal at Haifa .HaMifratz Port (Stage A)
• Final Geotechnical Interpretative Report- February 2012.
• Section III Technical Specs. Chapter 01.05 – Dredging, Reclamation and Disposal of Dredged Material. 2013/14.
2. DHI-Danish Hydraulic Institute, Horsholm, Denmark. New Container Terminal in Haifa- HaMifratz Port (Stage A).
• Examining the Environmental Impact of Shallow Water Dredging of Sand –Final Report, Volumes I and II- Environmental Impact Assessment. March and May 2013.
• EMP-Environmental Management Plan for the Dredging Plan- Technical Specs. Section III Annex 01.05A. 2013/14.
3. CAMERI Coastal and Marine Engineering Research Institute. Technion City-Haifa, Israel.
• Assessment of Environmental Aspects Related to a New Container Terminal (Haifa Bay Port) Layout A2 Final Report .P.N. 717/10 - December 2010.
• Artificial Sand Nourishment Downstream New Container Terminal (Haifa Bay Port) Layout A2-Final Report P.N. 718/10 -.December 2010.
• Changes in Sediment Transport Patterns due to the Different Size and Depth of the Sand Borrow Area Report P.N. 721/10. January 2011.
• Haifa Bay Port - Sedimentological Monitoring Study. P.N.847/17. November 2017.
4. Fugro West Inc. New Marine Container Terminal at Haifa. Design Ground Motion Criteria. DNHMCT–G.21 October 2009.

ENHANCING WORKABILITY BY OPERATIONAL WAVE MODELLING

by

François Dekeuleneer [1], Joris de Vroom[2] and Arjan Mol[1]

ABSTRACT

When dredging in exposed waters, wave conditions may seriously impact the workability of a dredging project. Especially stationary dredging equipment that makes use of spuds in order to remain in position and transfer the dredging forces to the seabed, like a backhoe dredger or a cutter suction dredger, is vulnerable for harsh wave conditions. The workability of such vessels is not only affected by the wave height, but also the wave period. Other types of marine operations, such as the construction of jetties, installation of wind turbines or the placement of scour protections are affected as well in their workability by the ambient conditions at sea.

Various regions all over the world are known for their problematic wave climate; the African west coast, the French and Spanish Atlantic coast, the Indian coastal waters, etc. are known for their long swell coming with long wave periods. But also less swell-dominated seas such as the North Sea may have severe wind sea systems with typical wave peak periods around 6 to 7 seconds. In extreme cases, even for large cutters, workabilities of less than 50% are not exceptional. Given the large stand-by costs of such specialized vessels, this can have a huge impact on the cost of a dredging project.

There is not only a considerable cost impact. Also the safety of the crew working on board of vessels in harsh conditions is at stake. Usually it is the responsibility of the captain to decide when the works need to be ceased in case of upcoming bad weather conditions. Therefore the captain needs to have a thorough knowledge of the limits of the vessel in terms of metocean conditions, and he/she should also have good insight in the current and upcoming weather conditions. When there is uncertainty in one of those elements, the captain's decision might be subjective and lead to unsafe situations or inefficiency:

- Unsafe working conditions follow from the fact that the equipment is being exposed to conditions beyond its workable limits. This could lead to damage to the equipment, for example damage to the spud, and uncontrollable motions of the vessel. In such case there is a risk for unsafe situations for the crew.

- Loss of efficiency is caused by a captain's decision not to work, while in reality the weather conditions are below the critical limits. This often happens after a period of bad weather, and conditions start to improve again, but the decision to resume the works is dominated by over-conservatism. The quality of the consulted weather forecasts also plays an important role in this process.

---

[1] DEME-group, Zwijndrecht, Belgium. De.Keuleneer.Francois@deme-group.com
[2] BMT Argoss, Amersfoort, The Netherlands. Joris.devroom@bmtglobal.com
[3] DEME-group, Zwijndrecht, Belgium. Mol.arjan@deme-group.com

In order to improve this situation, DEME has developed an operational tool in cooperation with BMT Argoss which aims to provide the on-board crew and site staff with information on the present and near-future sea states and whether operational thresholds are expected to be exceeded. The sea state is broken down in systems of common meteorological genesis which are considered to be statistically uncorrelated. With the use of response amplitude operators, key operators are determined and presented via a web application. Whenever the actual wave conditions are getting too rough the system will indicate that the workability limits are being reached and work should be ceased. Real time sea state data can be acquired from buoys that are deployed near the works. Future sea states are provided by a combination of operational atmospheric and wave models that typically deliver a five to eight day forecast window. To be able to further increase the accuracy and skill of these forecasts, the models are calibrated on the measured waves. The wave forecasts make it possible to plan the works more efficiently and to optimally use available workable windows. It generally results in less downtime, less damage and a safer working environment.

The tool has initially been applied for the Wheatstone downstream project, Australia, where a large access channel for a new port had to be dredged. This paper will discuss two other project cases where the tool has been applied; first the dredging works for a new highway at La Reunion island (Indian Ocean) and secondly the dredging and offshore installation works for the offshore wind farm Rentel in Belgium. The focus will be on the quality of the wave forecasts by operational wave models and the continuous calibration efforts which are increasing their reliability during the project.

1. INTRODUCTION

The Workability Tool (WoTo) has been developed to objectify the decision-making process regarding the weather conditions when active in harsh weather conditions. Combining measured and forecasted metocean parameters with the workable limits of vessels, the crew is able to evaluate the workability at any time. The comparison between the measured and forecasted wave parameters gives an appreciation of the confidence one can have in the forecasts.

The deployment of this tool has significantly increased the safety, the awareness, the efficiency and the planning on the different projects where it has been deployed. The reliability of the tool is highly dependent on the quality of the weather forecasts and, depending on the location of a project, the wave modelling can be more or less accurate. The development and improvement of wave models are therefore generally necessary to increase the reliability in the forecasts. These step by step improvements are then validated and/or calibrated using measured metocean data.

After a short introduction of the different hardware and software components, two case studies will be presented and discussed.

2. DESCRIPTION OF THE WOTO

The workability tool needs different entries to compute different outputs. Some inputs and outputs of the tool will be briefly explained below.

2.1 Measured Wave Parameters

To evaluate the current wave conditions, statistical parameters have to be derived from wave energy spectra. These can be obtained from different instruments (radar, buoy, ADCP, etc.). The focus will here be on directional wave rider buoys.

Most wave rider buoys are generally delivering a wave spectrum every 30 minutes. To get closer to a real-time monitoring of the waves, specific software has been developed to reprocess the wave raw displacements data and obtain new wave spectra containing the buoy displacements of the past 30 minutes every 3.75 minutes using a rolling buffer.

To obtain the wave statistical parameters from the spectra, a splitter has been developed by BMT Argoss and integrated in the WoTo. The splitter uses a wind component (from the forecast or measured) to allocate the spectral energy to a swell or a wind system.

A significant wave height (Hs), a mean period (Tm) and a direction (Dm) will thus be obtained for both the sea/wind and swell components. The system will then look at which system is the most impacting the workability of the vessel.

2.2 Forecasted Wave Parameters

The forecasting models are run several times a day by BMT Argoss and the output (Hs, Tm and Dm for both sea/wind and swell components) is imported in the WoTo. Depending on the operations, the frequency of the forecasts is adapted. In the WoTo the forecast is shown next to the measured data (Figure 1) in order to get a feeling on the quality of the forecast.

The model setup will be discussed in more detail in the different cases.

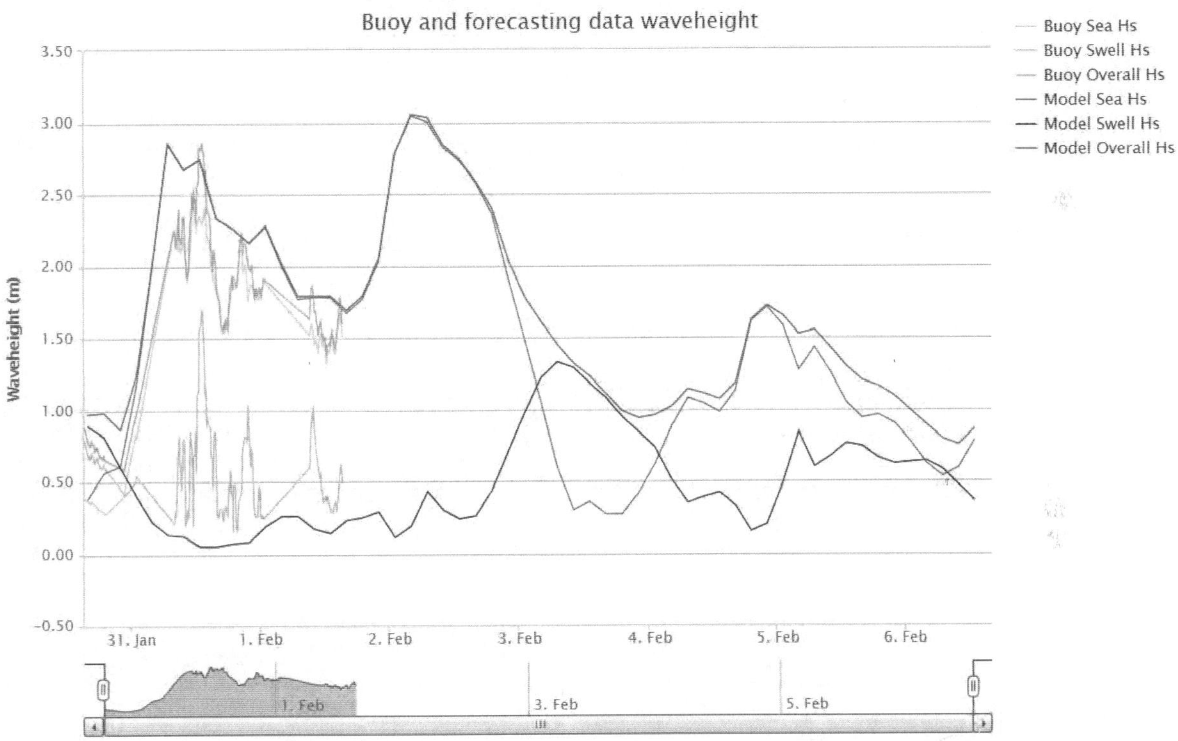

Figure 1: visualization of the measured and forecasted statistical wave parameters

2.3 Workable limits of the operation

Depending on the operation of interest, different workable limits are considered. Stationary vessels will generally be modelled in a diffraction model to determine the forces/movements generated under specific wave conditions. Each vessel and activity have their limits and on their basis, workability tables are generated. The WoTo shows the actual wave conditions in a workability plot from which it becomes clear if the conditions are workable or not (Figure 2).

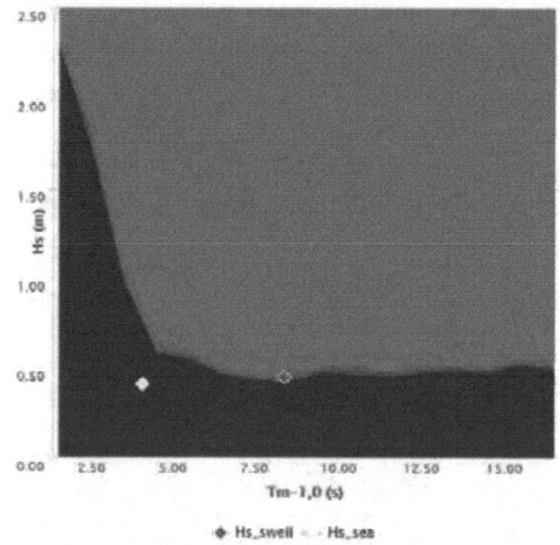

Parameter (Buoys-BUOY2)			
	Hs (m)	Tm-1,0 (s)	Direction (deg)
Swell	0.45	8.42	24.1287
Sea	0.41	4.09	58.2697
Overall	0.60	6.45	40.2927

Figure 2: visualization of the comparison between the measured waves and the workable limits of an activity.

3. PROJECT CASES

3.1 Rentel

Project Description
Rentel is developing an offshore wind farm in the Belgian territorial water in a zone designated for offshore energy (Figure 3). The wind farm development is located approximately 32 km offshore from the port entrance of Zeebrugge and 42km from Ostend, in between the Northwind and C-Power concessions. The wind farm will consist of 42 wind turbines generating a nominal power of 309MW.

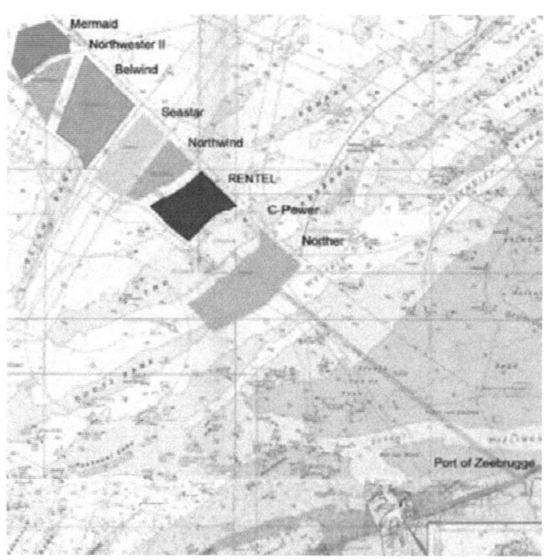

Figure 3: Overview of the Rentel wind farm concession southeast of the Lodewijk Bank and northwest of the Thornton Bank

Dredging International NV (part of the DEME-group) is the main contractor for the works and is further supported by other companies of the DEME-group as shown in Figure 4. The contract involves various operations such as the installation of the foundation and transition pieces, the cable laying and burial, the scour protection, installation of the turbines, etc. Figure 5 shows one of the vessels Geosea is using for the installation of the foundations.

Figure 4: Rentel contractor organisational structure.

Figure 5: Geosea's vessel Innovation arriving on a windfarm installation site.

Climate

The southern North Sea has a typically moderate climate: unsettled conditions prevail and strong temperature variations do hardly occur. This is caused by the predominant (south-) westerly airflow, which is found in between the semi stationary Azores high pressure area and low pressure areas, which occur frequently over the northern Atlantic Ocean and usually track to the East or North East, close to Iceland towards (northern) Scandinavia.

The activity of the low pressure areas varies throughout the year. Strongest low pressure areas are found in the second half of autumn and during winter, but also in other times of year active low pressure areas may affect the North Sea and cause strong winds and high waves. Strongest winds and highest waves are however often caused by relatively short lived so called 'secondary lows', which develop in a frontal system over the British Isles or North Sea and track rapidly to the east or northeast. After the passage of a major low pressure area to the north of the North Sea, associated high swells often enter the North Sea from the NNW. These swells usually reach to the Rentel location. On its way to the south however, wave height usually decreases significantly before arriving at the Rentel location.

At times the Azores high pressure area extends to the east or northeast. Depending on the exact position, this may result in periods of calm conditions over the North Sea. These calm interruptions can vary strongly in duration (between parts of a day till weeks) and occur year round, but are most frequent during spring and summer. The more northerly position of the Azores high pressure area during spring and summer causes the predominant wind direction being more from the W or NW by then. These westerly and north-westerly winds often bring clear skies and good visibility. But as the sea surface temperature is relatively low in spring and early summer (compared to the air over the region), fog fields develop occasionally during warm periods and may cause low visibility, even for several days.

From time to time the Azores high pressure area extends a strong ridge over the British Isles and/or Scandinavia. When positioned over Scandinavia, these high pressure areas cause the wind mainly to blow from easterly directions over the North Sea. These winds can be pretty strong (7-8 Bft) in winter, but wave height at the Rentel locations is most of the time limited due to the sheltering by the Dutch coast. The easterly winds typically bring good visibility since the advected air is usually dry. If the high pressure is located over the British Isles, winds will be predominantly blowing from N-NW directions. Wind speed over the North Sea is often moderate or fresh during these events. High pressure areas over Scandinavia and the British Isles occur most frequently in spring.

Figure 6 shows directional roses of 10 meter wind speed (1-hour sustained) and significant wave height from WaveWatch III Eushelf grid point 51°40'N, 3°00'E. The wave rose shows the two major wave systems from the north and from the southwest. Examples of WW3-eushelf model output corresponding to these wave systems are shown in Figure 7 and Figure 8.

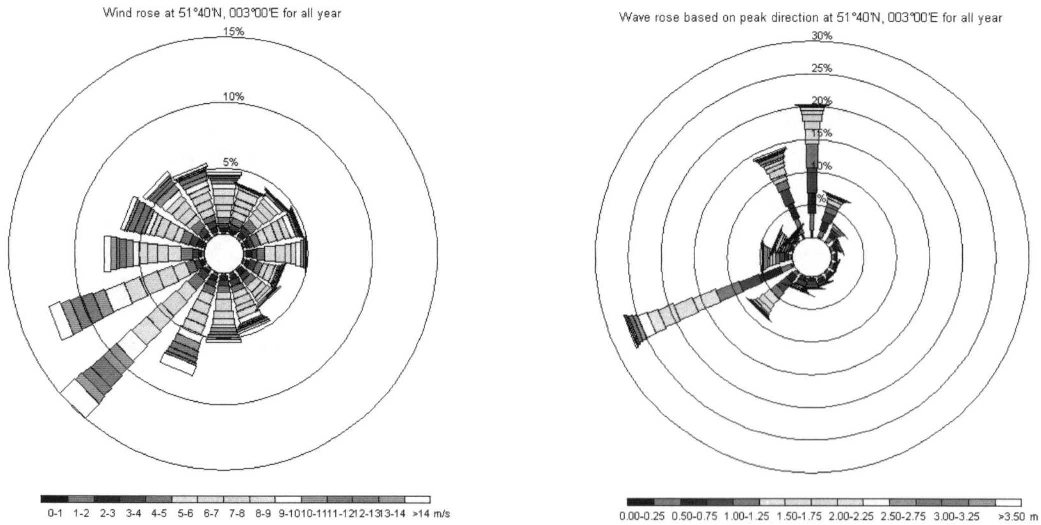

Figure 6: Directional roses of u10 (left) and Hs (right). Data was taken from grid point 51°40'N, 3°00'E from regional WW3-eushelf model. Years used 2016-2017.

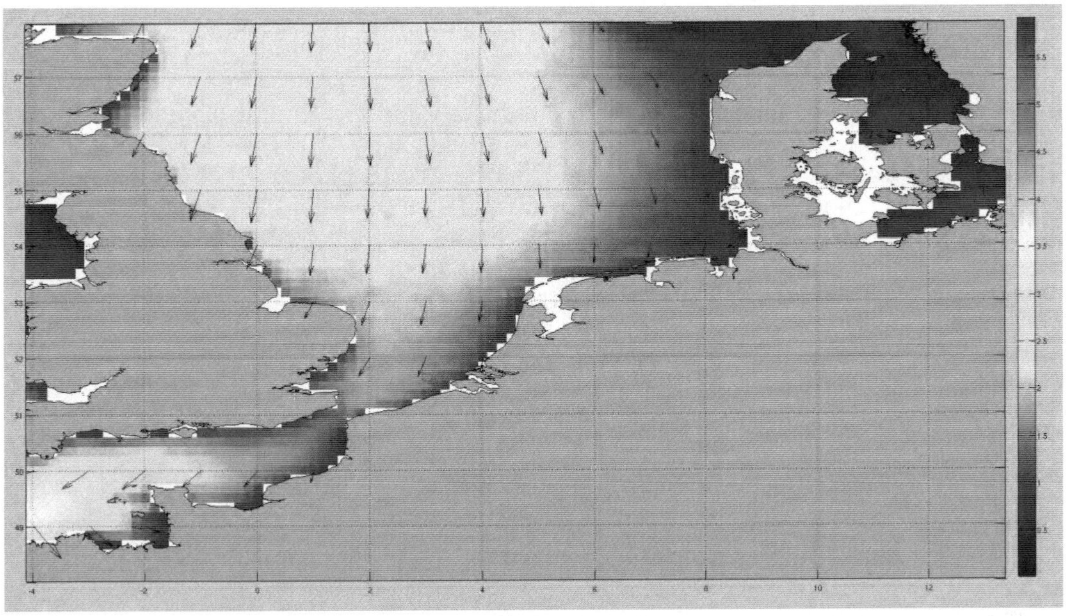

Figure 1: Example of WW3-eushelf waves from northern direction at Rentel (07-jan-2018 03 UTC). project location is shown by the pink square.

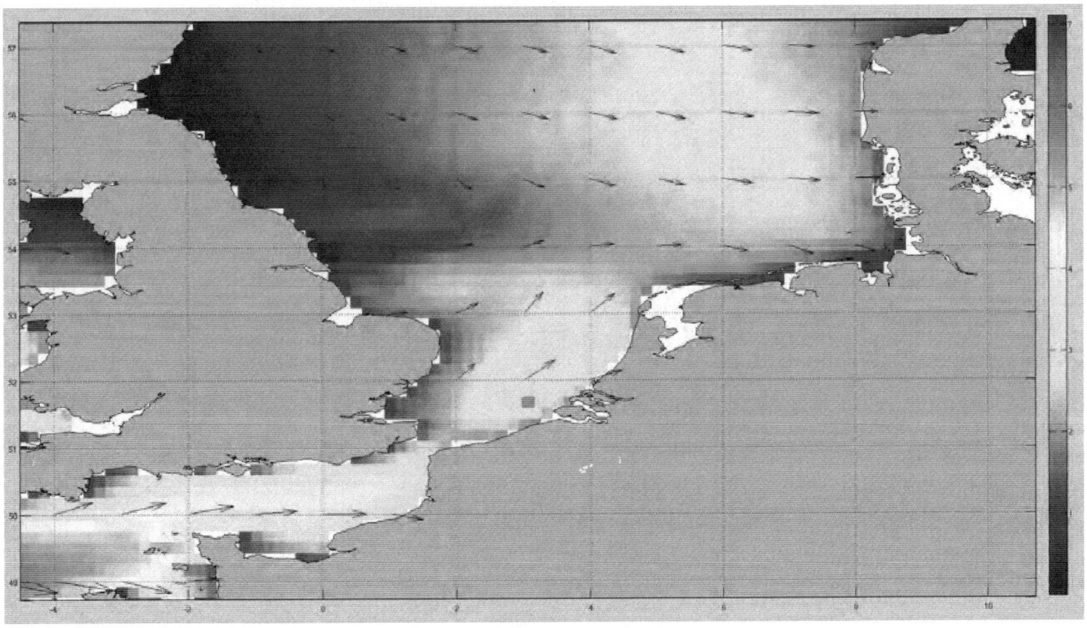

Figure 2: Example of WW3-eushelf waves from southwestern direction at Rentel (18-jan-2018 at 06 UTC). Project location is shown by the pink square.

Model Description

In the southern North Sea, BMT Argoss operates two regional wave model grids. Specifications of these grids are shown in Table 1 and outlines are shown in Figure 9 and Figure 10.

The larger domain covering the European continental shelf (Figure 9) is the regional WaveWatch III EU-Shelf grid with spatial resolution of 10 minutes (~17 km). It is a 3rd generation wave model based on the WaveWatch III code. The WaveWatch model is particularly suitable for propagating offshore waves. The wave model is driven by GFS wind data provided by the marine modelling branch of the National Center of Environmental Prediction (NCEP). It receives boundary conditions from the WaveWatch III Global wave model.

The smaller domain covering the southern North Sea and the English Channel (Figure 10) is the local SWAN DunkA grid with spatial resolution of 6 minutes (~10 km). The SWAN model is a third generation spectral model developed at Delft University of Technology, that computes random, short-crested wind-generated waves in coastal regions and inland waters. The model is based on the wave action balance equation with sources and sinks. SWAN accounts for the following processes:

- Wave propagation in time and space, shoaling, refraction due to current and depth,
- Frequency shifting due to currents and non-stationary depth.
- Wave generation by wind.
- Three- and four-wave interactions.
- Whitecapping, bottom friction and depth-induced breaking.
- Dissipation due to aquatic vegetation, turbulent flow and viscous fluid mud.
- Wave-induced set-up (not applied in this project).
- Transmission through and reflection (specular and diffuse) against obstacles

The wave model is driven by WRF winds obtained from BMT Argoss local WRF nweurope with spatial resolution of 27 km.

Grid	Engine	Forcing (resolution)	boundary	Outline ([lon],[lat])	Resolution
WW3-Eushelf	WaveWatchIII	GFS (25 km)	WW3-Global	[-15, 31], [40, 66]	17km
SWAN dunkA	SWAN	WRF nweurope (27 km)	WW3-eushelf	[0.17, 4.46], [49.67,52.10]	10 km

Table 1: Model specification for Rentel

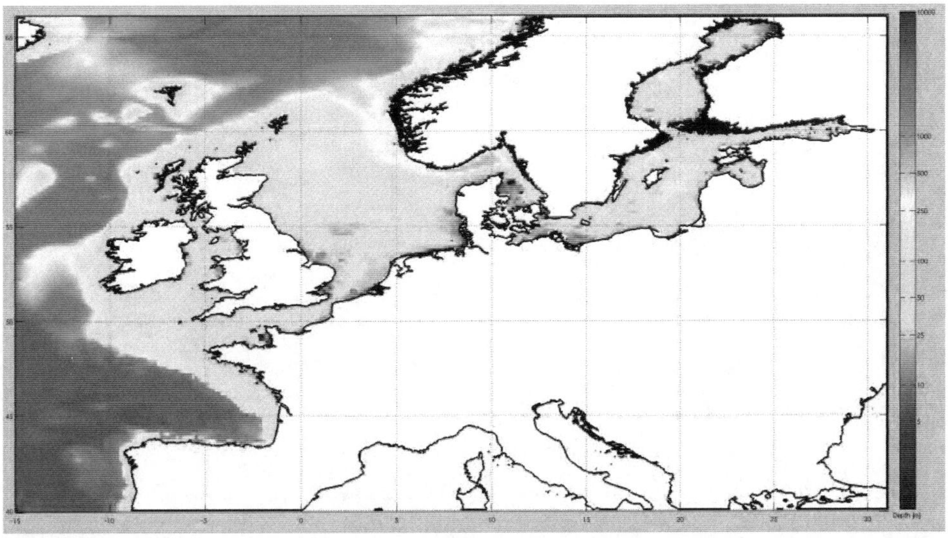

Figure 9: WW3-eushelf model grid outline and depth. Project location is shown by the pink square.

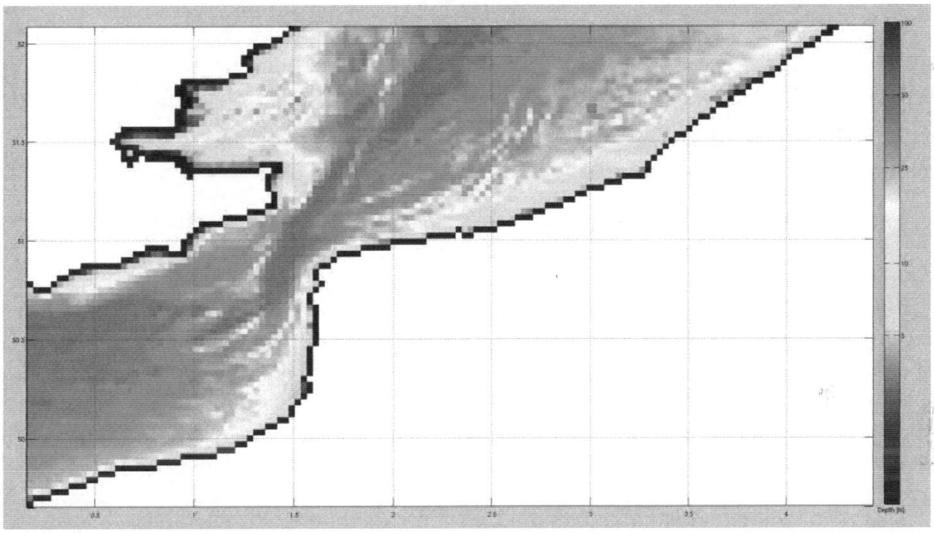

Figure 10: SWAN-dunkA model grid outline and depth. Project location is shown by the pink square.

3.2 La Réunion

Project Description

La Réunion is a French island and department in the Indian Ocean 700km east of Madagascar. On this island a new highway is planned from the capital Saint Denis to La Possession. This 9km long highway will replace the existing cliff road. This road, constructed in 1976, has two lanes in both directions and was designed for 10,000 vehicles per day. Nowadays about 60,000 vehicles make use of the road every day resulting in a lot of accidents and major congestions. Further, the existing road is subject to rock-falls from the steep cliffs adjacent to it and flooding during tropical storms resulting in unacceptable traffic jams. During heavy rain storms the road is even closed for any traffic.

The new highway will be partly constructed as a viaduct and partly as a causeway (see Figure 11) and it is designed to be operational during wind speeds up to 150 km/hr and waves up to 10 m. The project is financed partly by European funds. The main contractor for this project is a joint venture between French companies. SDI (DEME-group) is working in this project as a subcontractor.

Figure 11: Artist impression of the new highway which will be partly constructed as viaduct and partly as causeway.

The work consists of dredging a trench and making the foundation of the revetment for the causeway parts of the highway. Three sections of the highway will be constructed as a causeway. The total length of dredged trench is about 3,070 m. The trench has a width of 25 m at the base and a depth of about 5 m below existing seabed. The total dredging volume amounts about 250,000 m^3. The dredging works are carried out by the backhoe dredger Pinocchio. The dimensions of this dredger are 60x19 m^2. It has three spuds with a length of 40 m and the total installed power is 2,416 kW. Two methods of disposal of the dredger material are used. The first method is to sidecast the material using a pontoon which is moored alongside of the Pinocchio. A second method that has been applied is to use 850 m^3 barges to dispose the material. The dredger material is partly reused in the body of the revetment.

Apart from the dredging works, the work consists of placing of two layers of rock. The first layer is a 2-150mm rock filter of which a total of 170,000 tons must be placed. The second rock layer consists of 0.2-1 ton rock with a total weight of 340,000 tons.

Climate

The volcanic island La Réunion is characterized by a complex orography. High peaks and deep valleys have a strong effect on the atmosphere over and around the island. In winter and spring, the trade winds blow over the region. Strong lee effects can be observed on the west side of the island which are highly correlated to the background wind conditions over the region; small differences in the background wind direction cause significant changes in wind conditions at La Possession. For example: with winds blowing from the E or ENE, the north-easterly wind velocity is higher, up to 20-25 knots, and blows throughout most of the day. Under conditions with a slightly veered background wind (ESE) the enhanced north-easterlies are sparser and weaker. They are then alternated by weak sea breezes. Sea breezes play a very important role in the wind climate on and around La Réunion. A background sea breeze is caused by the difference of surface temperature between land and sea.

Diurnal heating of the island surface by the sun causes, in combination with the high mountains on the island, a strong convection over the island in general and particularly over the mountain tops. This phenomenon occurs virtually every afternoon, especially in summer. The opposite takes place in the evening and early nights. As the convection, triggered by the heating of the sun, stops at the end of the day, the air in the mountains starts to cool. The colder air flows downhill and reaches the coastal areas as a weak to moderate land breeze. When the temperature drop in the mountains is relatively strong in comparison to the one over the coastal areas, the air in the mountains will start to descend rapidly. Like any fluid it will follow the easiest path which is through valleys towards the coast. These so called katabatic winds can reach high speeds and occur very locally. Occasionally, these gusty winds can last for approximately (half) an hour.

Waves do not grow significantly due to the diurnal wind effects at the site of interest as the forcing surface winds are so short lived. Most wave energy consists of swells from the Northern Indian Ocean. During the stronger trade winds in winter waves from the ENE can refract around Saint Denis which results in waves travelling along the shore and reach La Possession. Occasionally, a cross-sea is found with the additional wave system generated between La Réunion and Madagascar. During summer low, variable winds are, as an exception, alternated with severe storms (CEULEMANS & HULST, 2016).

Figure 12 shows directional roses of 10 meter wind speed (1-hour sustained) and significant wave height from WaveWatch III Global grid point 22°00'N, 3°00'E. The wave rose shows the two major wave systems from the north and from the southwest. Both wave systems curve around the Island of La Reunion and merge on the northern side of the island. This results in a sea state in which dominant peak switches between both wave systems. This is illustrated by a directional rose of significant wave height and peak direction in Figure 13. The data was taken from the swan reunionD grid using model grid point at the project site 20°52'21.1S, 55°24'31.0E. A spatial impression of this process is shown in Figure 14. This shows significant wave height and direction from the reunionC grid. The project site (20°52'21.1S, 55°24'31.0E) is shown by the pink square.

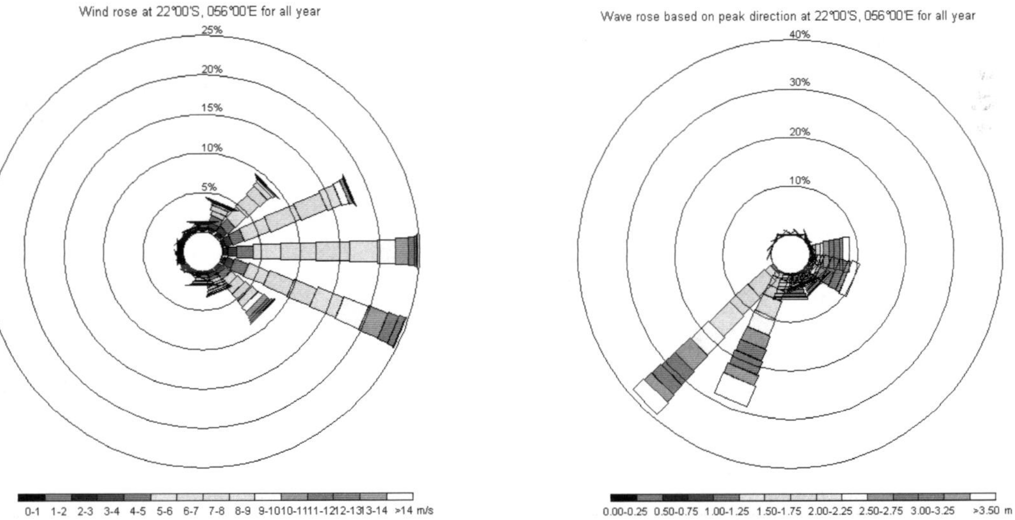

Figure 12: Directional roses of u10 (left) and Hs (right). Data was taken from grid point 22°00'S, 56°00'E from WW3-global model. Years used: 2016-2017.

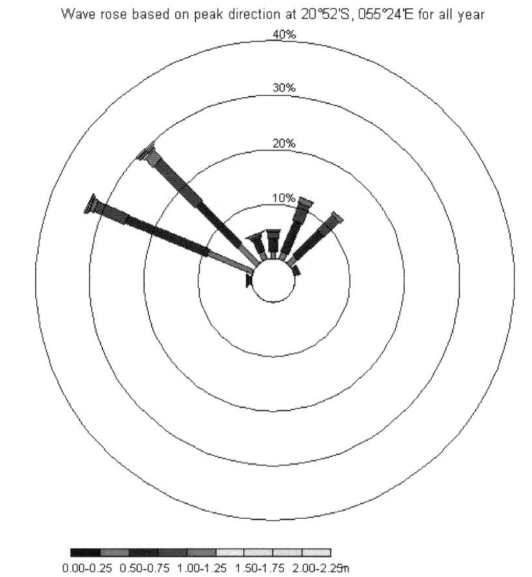

Figure 13: Directional rose Hs. Data was taken from grid point 20˚52'21.1S, 55˚24'31.0E from swan reunion model. Years used: 2015

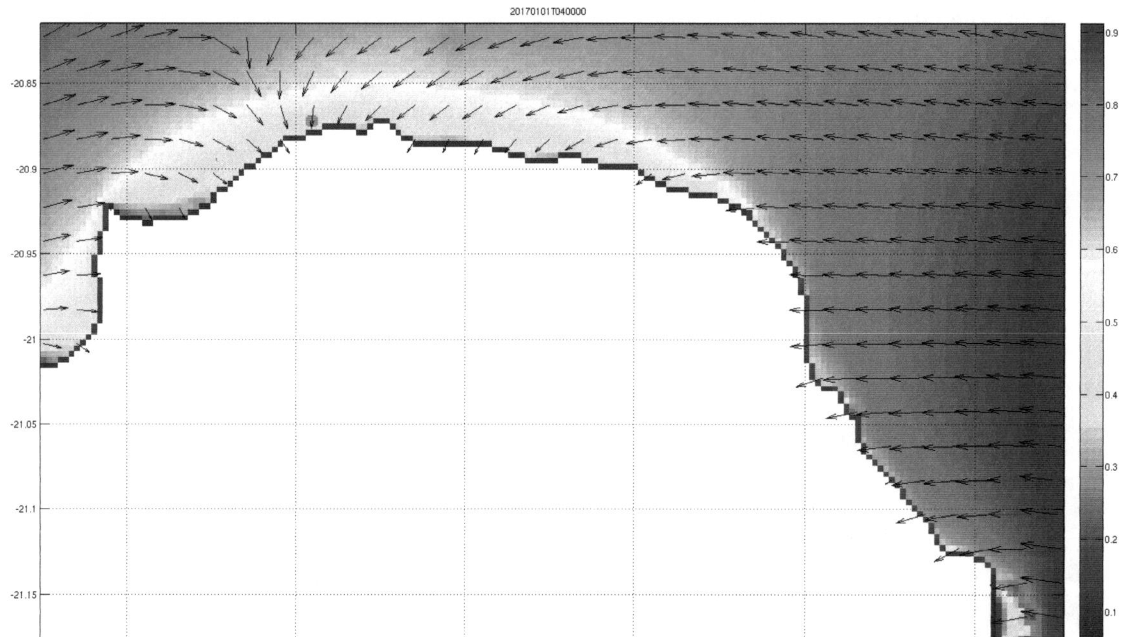

Figure 14: Significant wave height and direction from wave grid reunionC on 01-jan-2017 at 04:00 UTC. The project site is shown by the pink square

Model Description

The BMT Argoss model suite for La Reunion comprises of four nested SWAN grids. Specifications of these grids are shown in Table 2 and outline is shown in Figure 15 (please refer to Rentel model description for description of the SWAN modelling software). The large SWAN domain (reunionA) receives its boundary conditions from the WaveWatchIII Global wave model.

Grid	Engine	Forcing (resolution)	boundary	Outline ([lon],[lat])	Resolution
reunionA	SWAN	WRF-D01 (27 km)	WW3-Global	[55,56], [-21.5,-20.5]	10 km
reunionB	SWAN	WRF-D01 (27 km)	reunionA	[55.17,55.9], [-21.4,-20.8]	2km
reunionC	SWAN	WRF-D02 (9 km)	reunionB	[55.25,55.85], [-21.18,-20.82]	300 m
reunionD	SWAN	WRF-D02 (9 km)	reunionC	[55.34,55.43], [-20.93,-20.85]	80 m

Table 2: Model specification for La Reunion model suite

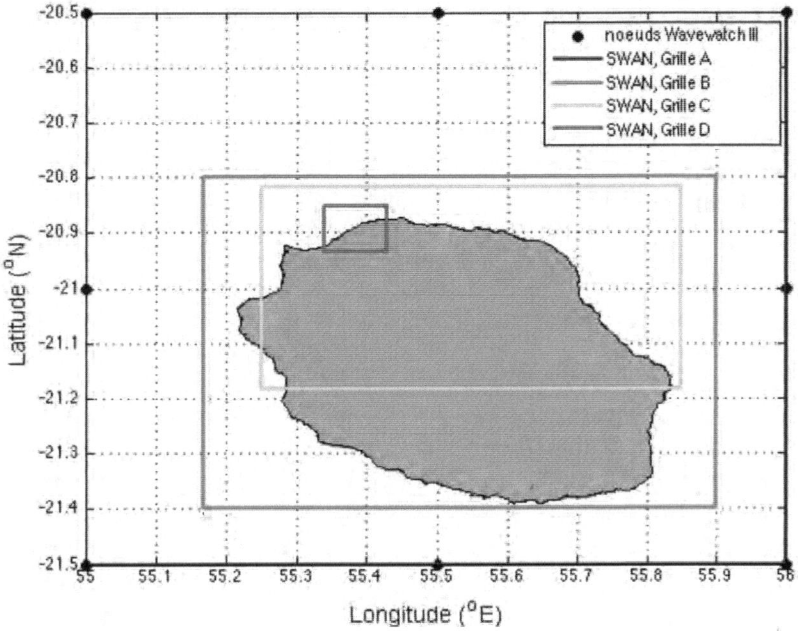

Figure 15: Extend of the swan grids for La Reunion

During the project, several model developments were tested. The developments that lead to improvement of model results are shown in Table 3.

In August 2017 we switched nesting the outer grid (reunionA) from 1D spectra (spectral wave information per frequency) to 2D spectra (spectral information per frequency and per direction). The assumption was that this would improve the wave direction since directional information is better represented in 2D spectra as compared to 1D spectra.

In October 2017 we switched off wave growth due to local wind. The local wind growth over grids reunionC and reunionD is assumed to be marginal, and it showed to effect the wave direction in a negative way.

Development	Description	Used in grid	Date in effect	Effects
Nest in 2d spectra	Nesting of reunionA grid in WW3-Global 2d spectra instead of 1d spectra to allow for better representation of directional information.	A	Aug 2017	wave direction
switch off windgrowth	Local windgrowth results in unrealistic mean wave directions from the N. Windgrowth switched off for grids C and D.	C and D	October 2017	wave direction, wave height

Table 3: Main model developments

4. RESULTS

The results obtained for both Rentel and La Réunion will be presented in a slightly different way. The North Sea being a well-known region where numerous offshore projects have been developed, the wave models are already performing quite well. The focus will thus be on the optimal use of these models by the meteorologists to provide manual forecasts getting as close as possible to the wave measurements.

On the opposite, the La Réunion region is a very particular environment where a lot of local phenomena influence the wave climate. The continuous monitoring of the waves combined with the know-how of BMT Argoss has led to a significant improvement in the accuracy of the forecasting. Being in the middle of the Indian Ocean, the project is greatly benefiting from this increased reliability.

In the sections below, the same type of figure is repeatedly used to indicate the performance of the model against the observations. These figures consist of an upper plot showing both the 0 to 24 hours, 24 to 48 hours and 48 to 72 hours forecasted parameter (blue colours) and the observed parameter (pink). The lower plot shows the bias where pink indicates forecasts higher than the observed data and blue indicates forecasts lower than observed.

4.1 Rentel

As explained in the model description here above, BMT Argoss is using both a SWAN and a WAVEWATCH III regional wave model in the North Sea. Comparing the model results to the observations, one realises that both models are not really accurate. The SWAN model tends to overestimate the wave height (Figure 16) whereas WAVEWATCH III is generally underestimating this parameter (Figure 17).

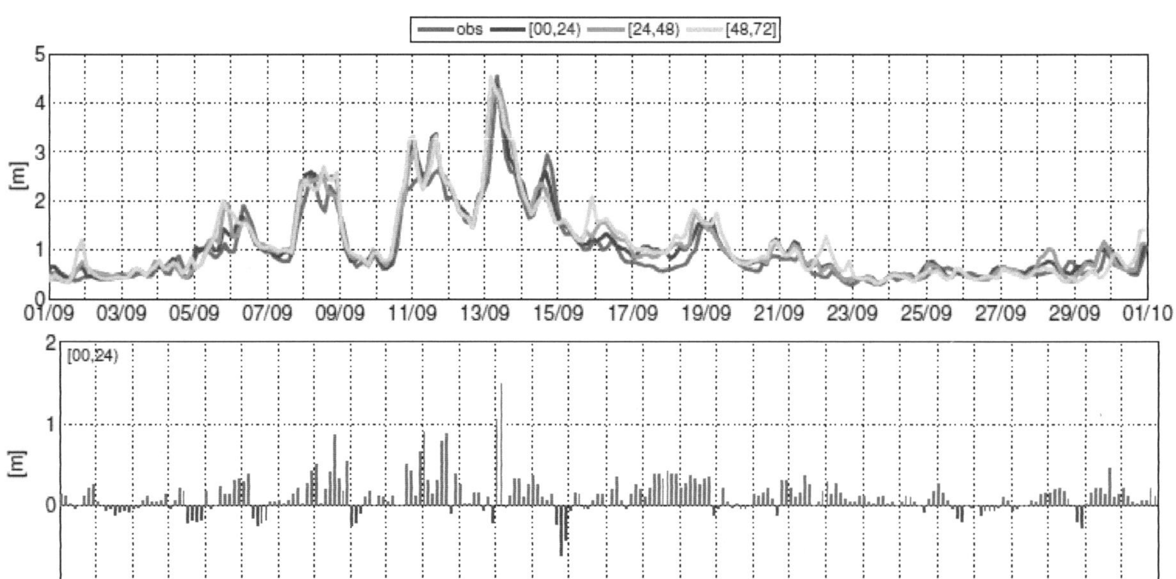

Figure 16: SWAN regional model significant wave height forecast for September 2017 (upper plot) and the bias (lower plot).

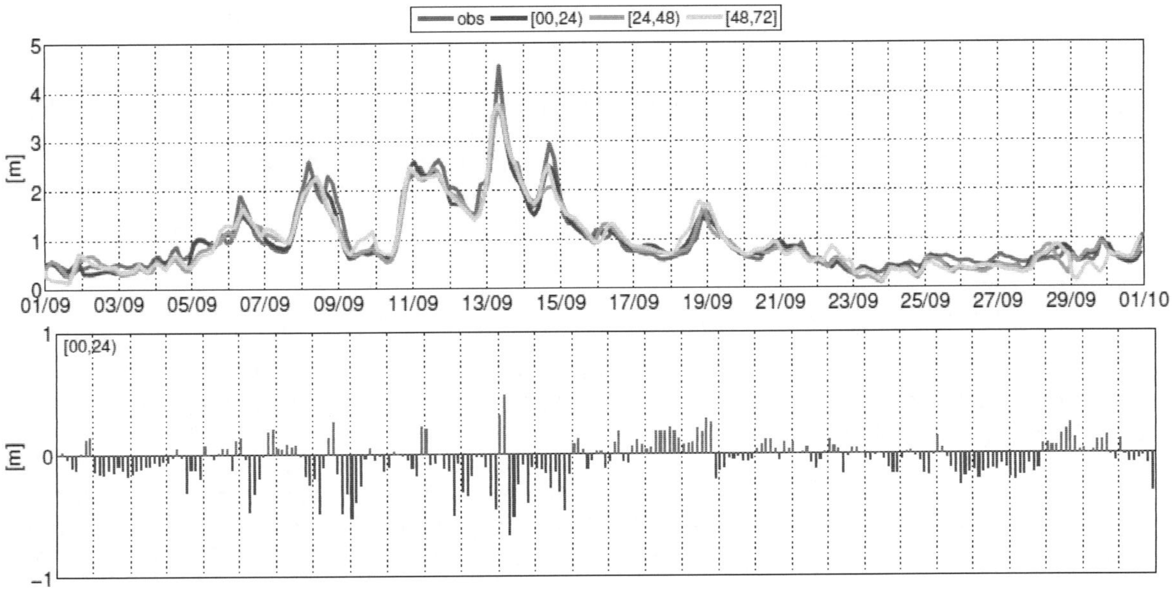

Figure 17: WAVEWATCH III regional model significant wave height prediction for September 2017 (upper plot) and the bias (lower plot).

Thanks to the comparison with the real-time monitoring, the meteorologist can use the output of both models and weigh them properly to offer the project team a constant reliable forecast (Figure 18).

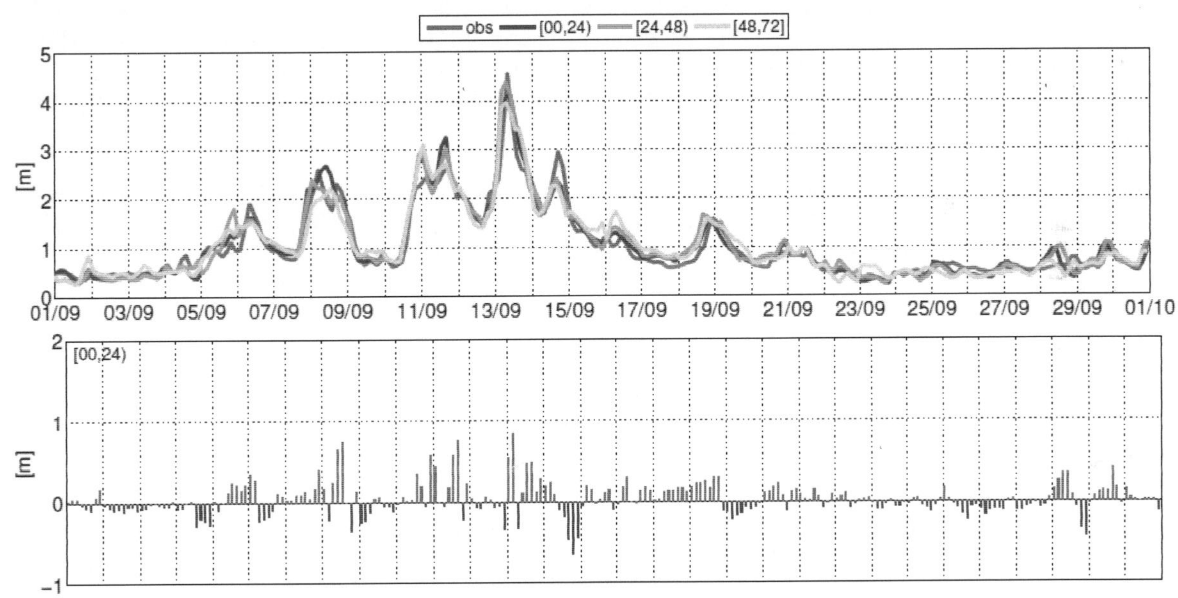

Figure 18: Manual significant wave height forecasts for September 2017 (upper plot) and the bias (lower plot).

Table 4 below compares the performance of both wave models (WaveWatchIII and SWAN) and also the manual forecast. The performance is quantified by the bias of the 0-24hrs predictions compared to the observations. From the table we see results consistent with the results in figures 16 till 18. SWAN overestimates wave height, WaveWatchIII underestimates wave height and the manual forecast is in between both models. Furthermore, the manual forecast shows to be slightly high compared to the observations. Apparently, the relatively high SWAN results affect the manual forecast. Based on this we conclude that giving more weight to WaveWatchIII EU-Shelf and giving less weight to SWAN should improve forecast results.

ME [00,24]	SWAN	WaveWatchIII	Manual forecast
Sig. wave height [m]	0.12	-0.05	0.04

Table 4: Comparison of bias for the forecast of September 2017 between WaveWatchIII model, SWAN model and manual forecast.

4.2 La Réunion

As explained in the section on the climate of La Réunion, the island is prone to very complex refraction as well as local phenomena. The model has thus needed some fine tuning to accurately predict the wave height and direction.

Due to these refraction and local effects, the forecast and measurement locations are of crucial importance. The results presented here will focus on one location to keep it short and clear for the reader. To follow the evolution of the works, the wave rider buoy and forecast locations are modified on a regular basis.

In order to demonstrate the impact of the model performances that have been implemented in August-October 2017 (Table 3), results from a period before the improvements (June 2017) have been compared to a period after the improvements (February 2018).

Note that the mean wave period is not further discussed here. It is not straightforward to compare mean wave periods from models to mean wave periods from buoys. This is due to the fact that the wave model and buoy may use different techniques to compute the mean wave period. Furthermore, the mean wave period is a parameter that on average over the entire globe is useful to help describe sea states, but may be less meaningful and useful in particular situations such as double peaked spectra.

Below, Figure 19 and 20 show comparisons of predicted and observed significant wave height and wave direction for the period of June 2017 (before model improvements). Figure 21 and 22 show the same figures for February 2018 (improved model).

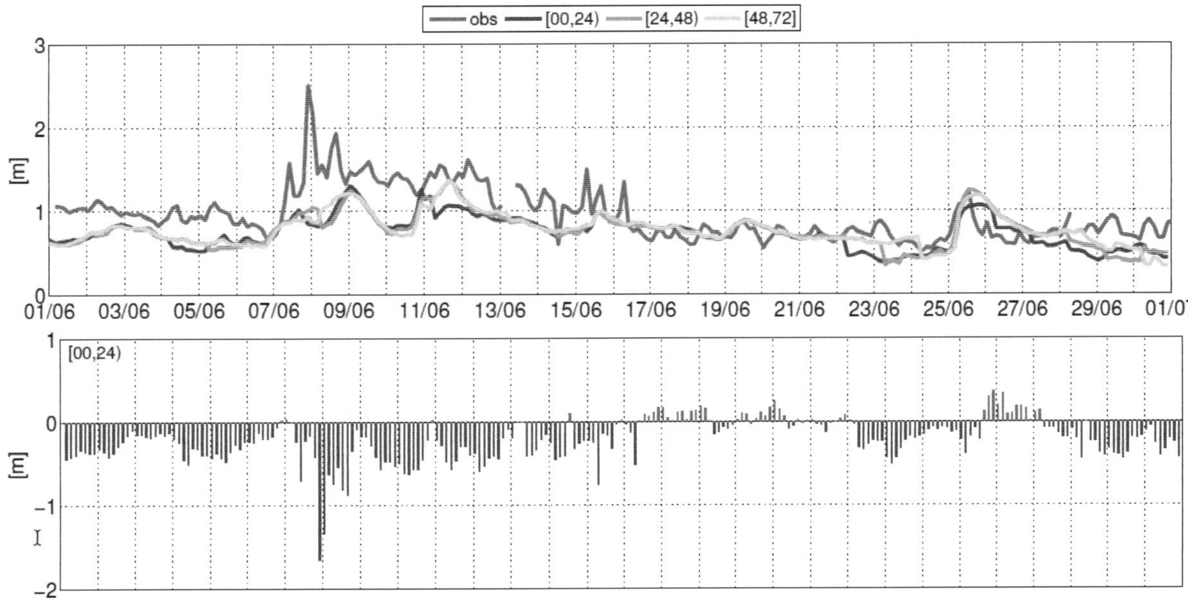

Figure 19: Wave height forecast for June 2017 (upper plot) and the bias (lower plot).

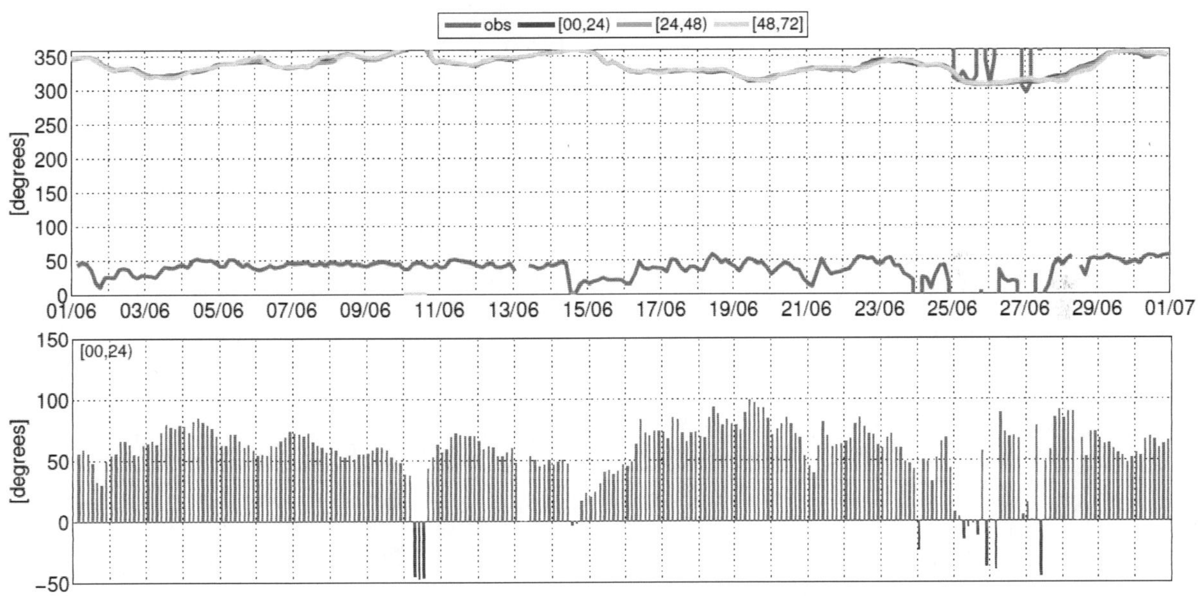

Figure 20: Wave direction forecast for June 2017 (upper plot) and the bias (lower plot).

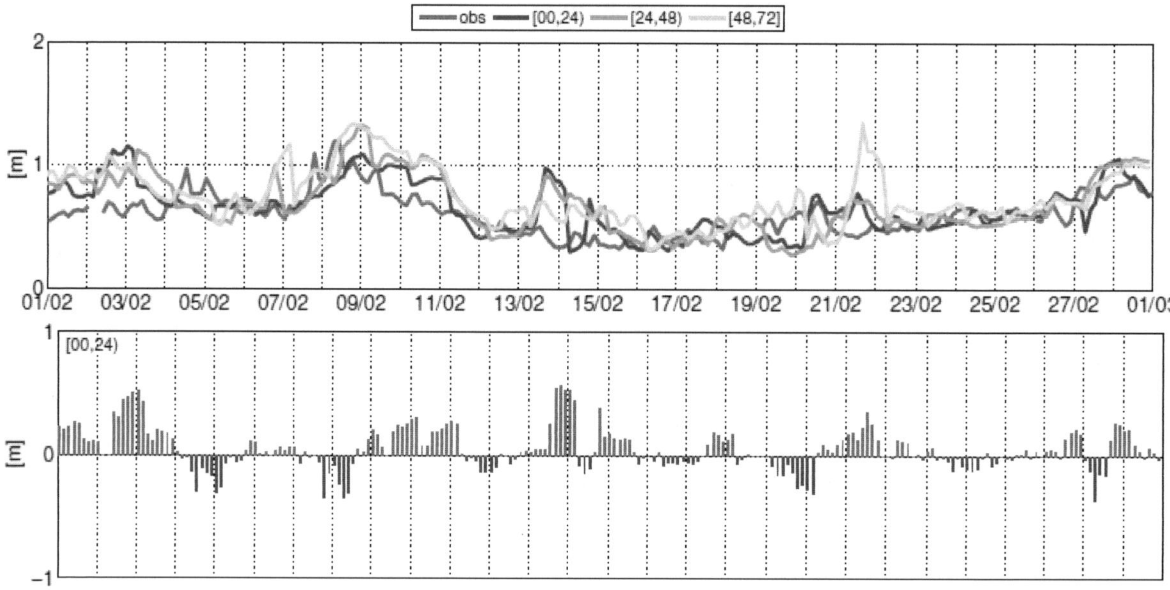

Figure 21: Wave height forecast for February 2018 (upper plot) and the bias (lower plot).

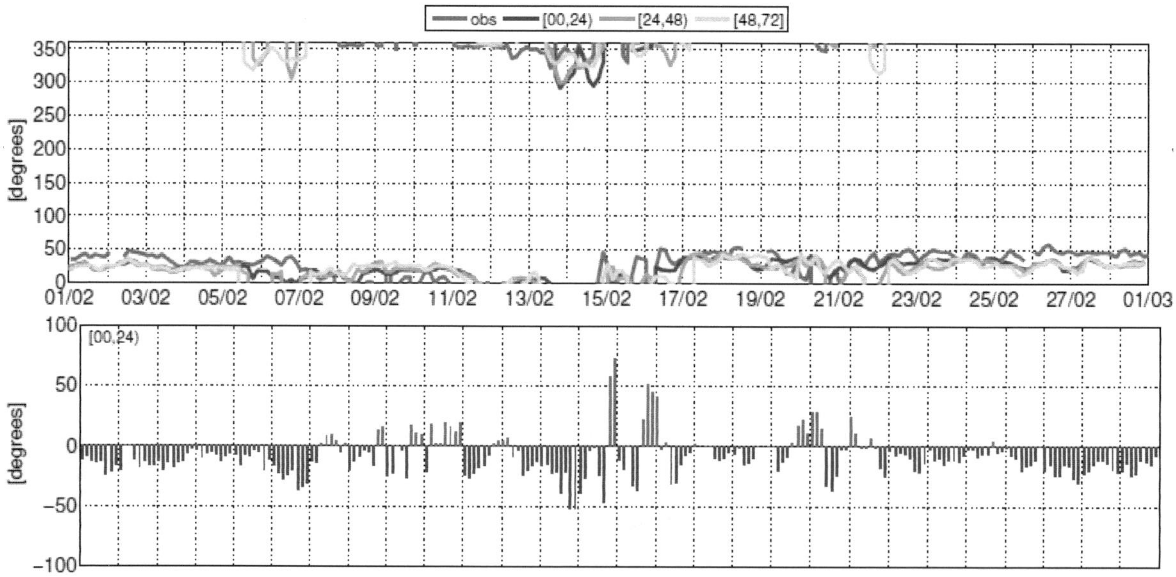

Figure 22: Wave direction forecast for February 2018 (upper plot) and the bias (lower plot).

Table 5 presents the bias of the model performance for a period before model improvements (based on April to June 2017) and after the model improvements (January-February 2018). These numbers and the figures 19 till 22 show that in the old situation, the forecast on average significantly underestimated the wave height at the project location and that the wave direction was systematically off by about 40 degrees (observations: northeast, model: north), while after improvements the wave height on average was only slightly overestimated while the wave direction is pretty much spot on.

ME [00,24]	Before model improvements April-June 2017	With improved model January-February 2018
Sig. wave height [m]	-0.24	0.08
Wave direction [deg]	37	-5

Table 5: Comparison of bias for the forecast before and after model improvements.

5. DISCUSSION

The two project cases above show the efforts being done to improve the predicted metocean conditions on site by using feedback (observations) from the site. There are some restrictions however that have learnt us that it is impossible to let the forecasts exactly match the observations:

- It is hard to find a consistent data set from buoy measurements on a project due to the fact that sometimes the buoy is displaced by the project crew, the buoys are taken out of the water (protection during storms), gaps in the data due to bad signal (remote sites), problems with acquisition software etc.
- It is not always straightforward to compare mean wave periods from models to mean wave periods from buoys. This is caused by the fact that the wave model and buoy may use different techniques to compute the mean wave period. Furthermore, the mean wave period is a parameter that on average over the entire globe is useful to help describe sea states, but may be less meaningful and useful in particular situations such as double peaked spectra.
- Comparing wave model output to buoy data can result in significant improvement of model results. However, there are a few limitations to this. These limitations include
    - Representability of model results for the buoy location
    - Wave model resolution
    - Differences in computation method of integrated parameters (wave height, wave period) between buoy and model
    - The wind data source driving the wave model. Poor quality wind data will in general result in poor quality wave data.
    - Quality of bathymetric data (for nearshore location)

6. CONCLUSIONS

Working in exposed marine environments may seriously affect the workability of the equipment being used in a project. Given the large stand-by costs of specialized offshore equipment, this can have a huge impact on the cost of those projects. There is not only a considerable cost impact. Also the safety of the crew working on board of vessels in harsh conditions is at stake.

The Workability Tool has been deployed to inform the crew and site staff with information on actual and upcoming wave conditions linked to the working limits of the equipment. The tool has proven itself in numerous situations in order to limit the downtime, work more safely and to better plan the works. Besides, the tool makes the bridge personnel aware of the metocean conditions. Based upon an increase of confidence (due to comparison wave forecast and real-time measured waves) the crew can now easily take an objective decision to stop or start or replace the works.

Input to the WOTO are forecasts coming from models. Wave models are developed to represent sea states on a global, regional and local scale. Many processes are included in the wave model, which are used all over the globe. Specific locations require specific tuning of the wave model parameters, such as wind drag, coefficients influencing refraction and others, and for this it is vital to have reliable buoy information. This paper has shown that the quality of the weather forecast can significantly improve during the project by making use of the field observations. The expert meteorologists are able to use the observations in the interpretation of the model forecasts to deliver an improved manual forecast. Besides, at set times they will use the observations to adjust and tune the wave models to better represent the observed conditions. When the quality of the forecast is increasing, the crew will also gain more confidence in the forecast which will help them to better plan the works and take responsible decisions.

References

Ceulemans, C., and Hulst, A.Th.C. (2016). Verification report La Reunion: Possession Monthly report, Amersfoort.

PLANNING, MANAGEMENT AND SUPERVISION OF DREDGING WORKS

by

Mauricio Torronteguy[1], Juliana M. Menegucci[2] and André Marques[3]

ABSTRACT

Brazil has 34 Public Maritime Ports distributed along about 7,500 km of coastline. In each of these Ports there are different terminals, some administered by public agencies, while others leased to private companies. Access to these terminals takes place through channels of different dimensions and intricacies to navigation whose competence to dredge, whether for maintenance or even deepening, pertains to the Federal Government or the local port authorities. Exceptions to this rule occur in the Private Terminals, located outside Public Ports jurisdiction and whose competence of dredging keep up to the administration part.

Significant advances have occurred in the last decade with foreign dredging companies enter in the Brazilian market, fact that enabled the implementation and maintenance of several navigable waterways in Brazilian Terminals. But if, on the one hand, legislation and even governmental programs made it feasible to carry out dredging works of greater magnitude and complexity in Brazilian jurisdictional waters, many works did not reach the proposed objectives due to the lack of planning before the dredging execution and precarious management and supervision of such works.

This article presents good practices of dredging control and management in maritime terminals that ensure the adequate technical, financial and environmental control of the works, and helping to reduce waste of public and private resources and minimizing risks.

1. INTRODUCTION

In 2007, Brazil instituted the National Dredging Program (PND), in which Federal Government resources were invested to carry out several dredging works in the Brazilian Public Maritime Ports. In addition to structuring a specific long-term program for dredging works, the PND facilitated the entry of foreign dredgers into Brazil, enabling large public and private works, and attempting to reduce the past and ongoing shortage of technologies and dredging companies in Brazil.

According to official data from the Brazilian Federal Government, during the first stage of the PND between 2008 and 2012, approximately 73.0 million cubic meters of sediment were dredged in 16 Brazilian Public Maritime Ports (SEP, 2015). In the second and last stage of the PND, expected to have occurred between 2012 and 2015, additional 44.0 million cubic meters of dredging were planned in 4 Brazilian Public Maritime Ports. There is no official study summarizing the results of the second stage of the program, but our estimates show that 50% of the expected amount was carried out by the end of 2017. Therefore, in a 10-year period (2008/2017), Brazil invested federal resources to dredge 95.0 million cubic meters.

---

[1] Oceanographer, M.Sc., Atlântico Sul Consultoria, Brazil, mauricio@atlsul.com.br
[2] Civil Engineer and oceanographer, Atlântico Sul Consultoria, Brazil, juliana@atlsul.com.br
[3] Civil Engineer, Atlântico Sul Consultoria, Brazil, andre@atlsul.com.br

It is important to note that in Brazil some port dredging was also carried out with resources from the local port authorities, such as the Port of Itajaí, Port of Imbituba and Port of Santos, the latter annually dredging its berths, and more recently, investing in port channel dredging (CODESP, 2018). These cases, also carried out with public resources, represented in the last decade, according to our estimates, something close to 15.0 million cubic meters.

Considering also the significant dredging, especially of deepening dredging carried out by private terminals, such as dredging of Port of Açu (BOSKALIS, 2015 and 2016), Thyssen Krupp CSA Terminal (DEME, 2017); Port of Tubarão (CEPEMAR, 2010), Piaçaguera Channel (TORRONTEGUY *et al, in press*), Port of Sudeste (DEME, 2015), Jurong Shipyard (CTA, 2009), EMBRAPORT (CARVALHO, 2013); Terminal of Ponta da Madeira (DEME, 2015), among others, it is probable that Brazil dredged with private investments about 155.0 millions cubic meters in the period between 2008 and 2017.

Table 1 (below) presents an estimate of the total volume dredged in public ports and private terminals in Brazil in the last 10 years. It shows that the total volume dredged has been of circa 265.0 million cubic meters, averaging 26.5 million meters per year. It should be noted that the estimate presented here is a preliminary attempt based on information in the public domain but should be used with caution because it is not an official data source.

Resources	Total Volume, in Million m^3	Perceptual
Federal Government	95.00	35.80%
Local Port Authorities	15.00	5.60%
Private Terminals	155.00	58.60%
Total	265.00	100.00%

Table 1: Total Volume dredged in Public Ports and Private Terminal between 2008 and 2017.

To contextualize the Brazilian case in the international scenario, we have done a comparison with the survey conducted by the International Association of Dredging Contractors (IADC), which estimates the size of the open dredging market worldwide, including port dredging and works for mining, energy, tourism, urban infrastructure, among others. It is important to note that IADC does not compute the dredging carried out in the US and China as these are closed markets to international competition. Based on the latest IADC report (2017), the average dredging of 26.5 million cubic meters per year in Brazilian ports would amount to something like 3.5% of the total amount dredged in 2016 around the world (excluding China and the USA).

Brazil's position is relevant in the international open market, especially because in this percentage we do not compute dredging in waterways and in mining, two sectors that demand important dredging in Brazil. Besides the numbers expressed above, another robust evidence of the Brazilian relevance in the maritime port dredging international market is the presence, through either construction or commercial representation, of many of the main world's dredgers contractors.

Despite all the above, the fact is that the National Dredging Program has been practically deactivated, and today there is no new government program to replace it. In fact, the main current debates on port dredging focus on the search for a model that allows dredging in Public Ports to become the responsibility of the private sector, which, due to the success that privately funded port dredging projects in recent years, has been responsible for more than half of the volume dredged in the last decade, as shown in Table 1. Another point that has motivated the privatization debate is the evidence of lack of planning and supervision in some public works as indicated in the opinions of the Brazilian's Account Court (TCU), the control team of the Brazilian State that supervises federal fund distribution.

2. PLANNING OF DREDGING WORKS

The success of a dredging work depends initially on good planning and this is only possible if it is supported by appropriate contracts to the complexity, deadlines and resources involved in the work, as well as a detailed design that should be contains at least the following information:

- Areas to be dredged;
- Volume to be excavated;
- Depths to be reached;
- Horizontal and vertical tolerances to be adopted;
- Disposal site for the dredged material;
- Equipment's to be used;
- Productivity to be obtained;
- Works schedule;
- Costs estimated;
- Environmental constraints such as sedimentation rates, hydrodynamic conditions, sediment characteristics, presence of rocks or obstacles, proximity to sensitive environments such as corals, etc.

Unfortunately, some managers underestimate the importance of a detailed design before contracting the dredging works and neglect that it is one of the most fundamental steps for the proper planning of the work. This situation became more frequent in Brazil with the creation, in 2011, of the so-called Differential Recruitment Regime (RDC), a modality of contract in which the public manager, based on a conceptual design, can contract all kinds of infrastructure works, including dredging, while the responsibility of the basic and detailed design development belongs to future job executor.

Surely, as the projects are being developed, doubts and uncertainties arise, hampering the progress of the work due to lack of information and, in some cases, leading to errors in quantitative estimates and consequently, delays, interruptions and substantial cost increases.

It is not the objective of this publication to discuss if RDC funded works presented or present problems, but rather to alert to the fact that presently in Brazil there is a modality of contract that allows the public manager, under justification of less bureaucracy, to adopt the RDC contract model despite its obvious risks. Thus, although the RDC served to execute several public dredging works in Brazil, it opened the possibility of these works being contracted with insufficient design elements.

Finally, it should be pointed that risk reduction for both contractors and contracted is one of the main objectives of a construction contract, with general and specific contractual models for each type of construction. In the case of dredging works, the Blue Book, published by Fédération Internationale des Ingénieurs-Conseils (FIDIC), presents the minimum clauses and standards to be adopted in specific contracts for dredging and reclamations works (MADDOCK & VANDENBERGHE, 2016). It points out the concern in minimizing risks in dredging works, suggesting inclusion of a specific contractual clause of "Defined Risks", which in turn can only be determined and duly expressed in the contracts if there are studies and previous projects with appropriate level of detail that subsidize the identification of risks.

3. VOLUME ESTIMATES AND CHARACTERIZATION OF MATERIAL TO BE DRADGED

As obvious as it would be to estimate satisfactorily the volumes to be dredged and to know the soil type to be removed, it is important to state that in many dredge works, this information is treated in a superficial manner.

To get the volumes of material to be dredged in maritime areas, the internationally recognized methodology is to carry out bathymetric surveys. In this respect, Brazil has many companies capable of performing such services with well-established standards, while adopting the specifications of the International Hydrographic Organization (IHO) for hydrographic surveys (DHN, 2017).

Despite the above favorable scenario, public ports and private terminals in Brazil seldom carry out continuous bathymetric surveys, which are often carried out in isolation without being part of a depth control program, usually executed in periods prior to work contracting and frequently being outdated at the beginning of the dredging, generating changes in the amount of the volume to be dredged and consequently changes in the prices and terms of the original contracts. Such situations are common in many Brazilian Terminals and could be avoided with a continuous and long-term program that would monitor the depths in waterways, generating a useful database for the updated calculations of volumes to be dredged and for studies of determination of the sedimentation rates, information fundamental to the correct maintenance of dredging planning.

Advances occurred in Brazil regarding the characterization of the material to be dredged from 2004 with the entry into force of a resolution of the National Council for the Environment (CONAMA), which required environmental licensing to assess the physical-chemical and ecotoxicological characteristics of the material to be dredged (MMA, 2012). The resolution is of an environmental nature and addresses aspects related to the granulometry of the material but does not deal with the geotechnical characteristics of the material to be dredged, such as hardness and resistance, fundamental information in the decision-making process regarding the required equipment and in getting estimates of their productivity. To determine this information, geotechnical surveys are necessary, while geophysical surveys are fundamental for the correct spatial distribution of these properties. These effective methods are frequently neglected by the contractors, often leading to overestimated prices by the dredgers, due to the risk imposed by the uncertainties, difficulties of excavation or, otherwise, opening for works stoppages due to initially underestimated difficulties.

4. MANAGEMENT AND SUPERVISION

The management and supervision of dredging works aims to ensure that the work is executed as foreseen in the detailed design and in the terms and prices established in contract. In addition, it aims to comply with the environmental licenses and authorizations in force and the reduction of any risks to the contractors, especially with regards to navigation and the environment.

The most effective way for proper management and supervision is the contractor making himself present at the construction sites. For this, it is recommended that specialized companies be contracted to maintain a technical team capable of checking and recording the progress of the works and to promote the necessary project adjustments when necessary. Thus, it is suggested that contractors adopt the following management and control procedures in dredging works:

- Allotment of land and board inspectors during the entire period of the work to record if all excavations and disposal of material are carried out as planned in the design and in the sites previously determined and authorized, pointing out and correcting any deviations;
- Allocation of designers throughout the execution of the work whom make the necessary changes in the designs as interferences arise that affect the progress of the services;
- Allocation of HSE technicians throughout the project execution period to ensure that activities are performed in accordance with HSE related legislation and maritime standards;
- Allocation of managers during the entire period of work execution that assess whether the expected productivity is effectively achieved and whether work is performed within the parameters and goals established in the contract;
- Monitoring and continuous registration of the navigation and positioning of the vessels and boats allocated in the work to ensure, above all, that the disposals are carried out in the areas duly licensed;
- Frequent monitoring of bathymetric conditions in the dredging areas to monitor the evolution of changes in depths and for the calculation of dredged volumes for the purposes of measurement and payment of the companies responsible for the execution of the works;

- Frequent bathymetric or topographic monitoring in the areas of material disposal to monitor changes in the bottom level conditions of these areas.

It is also recommended that the management company should work together with the contractor not only during the construction work, but also from the initial phases, such as the elaboration of the technical dredging requisition, as well as the hiring of the executing company, evaluating the following items in the dredgers before their actual contracting:

- Technical qualification of the company to carry out the dredging works;
- Previous experience in similar works;
- Construction methods presented in the proposals;
- Dredging equipment and proposed productivities including navigation and production control systems;
- Support estruture;
- Capacity of the technical team and Organization chart;
- Plan of preventive maintenance of equipment and eventual impacts on the execution schedule;
- Execution schedule, which should clearly establish the planned deadlines for mobilization, execution and demobilization;
- HSE Procedures and Policy;
- Ability to meet the internal standards of the contractor;
- Responsibilities and costs due to possible changes in the schedule of execution of the work, considering breaking of equipment, passage and maneuvers of ships; environmental licenses, replacement of equipment, alterations of projects, and others;
- Responsibilities and costs due to design or execution errors;
- Price and forms of measurement and payment;
- Criteria for termination or contractual additions because of: delay, design changes, changes in the type of dredged material, changes in the location of dredging or dumping areas, changes in input costs such as fuel; excessive paralyzes, environmental liabilities, and others.

5. FINAL CONSIDERATIONS

The port dredging market in Brazil has been promising in the last decade, mainly due to the opening of the sector to the major players in the international market and private investments in the period.

There are still uncertainties about the future of public investments in port dredging in Brazil, however, it is quite likely that such uncertainties will be remedied in the short term, and new investment programs or even new contract modalities will be idealized by permitting dredging on waterways of responsibility from the government.

Our recommendations on dredging engineering design, characterization of material to be dredged and estimates of volumes, as well as on the management and inspection procedures of works are common practices around the world (USACE, 2015) and that already were implanted in Brazilian works, including public works, but have been more frequent and really institutionalized by private contractors, which is very probably a determining factor for the good performance that the private dredging presented in the last decade in Brazil.

6. REFERENCES

BOSKALIS (2015). Port of Açu, Brazil - Dredging, Reclamation and Rip-Rap. Project Sheet.

BOSKALIS (2016). Dredging contract for oil terminal expansion in Brazil. Press Release. Available in: https://boskalis.com/press/press-releases-and-company-news/detail/boskalis-awarded-eur-120-million-dredging-contract-for-oil-terminal-expansion-in-brazil.html

CEPEMAR (2010). Relatório de Impacto Ambiental da dragagem de aprofundamento do Complexo Portuário de Tubarão. Relatório Técnico CPM RT 008/10.

CODESP – Companhia Docas do Estado de São Paulo (2018). Portal das Licitações. Available in: http://www.portodesantos.com.br/transparencia/transp_licitacoes.php

CTA (2009). Relatório de Impacto Ambiental do Estaleiro Jurong Aracruz.

DEME (2015). DEME Activity Report. 2014.

DEME (2017). DEME Presentation. Retos portuarios en el Perú. Available in: http://www.fita-npa.org/downloads/pastevents/peru/5.Presentacion%20DEME%20Lima%202017%20vs%20ESP.pdf

DHN – Diretoria de Hidrografia e Navegação (2017). NORMAN 25 – Normas da Autoridade Marítima para Levantamentos Hidrográficos.

FIBRIA (2016). Dragagem do acesso ao Canal do Tomba - Caravelas/BA. Plano de Trabalho 2016/2017.

IADC – International Association of Dredging Contractors (2017). Dredging in Figures 2016. Available in: https://www.iadc-dredging.com/ul/cms/fck-uploaded/documents/PDF%20Dredging%20in%20Figures/dredging-in-figures-2016.pdf

MADDOCK, T & VANDENBERGHE, M (2016). A Review of the FIDIC Blue Book (Second Edition). Terra et Aqua. Number 145.

MMA - Ministério do Meio Ambiente (2012). Resolução CONAMA Nº 454, de 01 de novembro de 2012.

SEP – Secretaria Especial dos Portos (2015a). Sistema Portuário Nacional. Available in: http://www.portosdobrasil.gov.br/assuntos-1/sistema-portuario-nacional

SEP – Secretaria Especial dos Portos (2015b). Programa Nacional de Dragagem – PND. Available in: http://www.portosdobrasil.gov.br/assuntos-1/pnd

TORRONTEGUY, M; MENEGUCCI, J; SARAIVA, A; CAMARA, F; DE CASTRO, N; MARQUES, A; CAPORALLI, A & COELHO *(in press)*. Piaçaguera Channel dredging case: Confined Aquatic Disposal - CAD as an alternative for the destination of sediments not available to the ocean disposal. Submitted to XXXIV PIANC World Congress.

USACE - U.S. Army Corps of Engineers. (2015). Dredging and Dredged Material Management. Manual EM 1110-2-5025.

SETTLEMENTS MONITORING ON SOIL IMPROVEMENT BY PRELOAD IN THE RECLAMATION AREA FOR A NEW PORT AT COSTA RICA CARIBBEAN SEA

by

***Esteban Mayorga*[1] *and Luis Millán*[2]**

ABSTRACT

Geotechnical monitoring is used more and more in engineering works. Successful monitoring requires, on one side, to understand which are the limit states conditions and principal variables to be measured, as well as instrumentation, data and comprehensive processing techniques to analyze and transform sensed data into diagnostics for decision-making.

The aim of this paper is present the settlements analysis based on geotechnical monitoring instrumentation for the preloads during construction of the reclamation for the new Container Terminal, in execution at Moin, Costa Rica.

The current phase of the project, comprises the construction of a 600 meters wharf, the dredging of 10 million cubic meters in the waterways and the reclamation of a 30 hectares island required for the operations areas, including storage yards, buildings and other facilities and utilities (*Figure 1*). Due to fat clay layer on the soil profile, if no geotechnical improvement was done, high settlements on surface could be expected.

Figure 1: Location of the Moin Container Terminal (MCT).

Prefabricated vertical drains (PVDs) were installed all over the reclamation, which was then preloaded with several surcharges using the same dredged material. This was done to produce most of the expected total settlement, before the construction of the yard and structures, reducing settlements during the operation phase, maintaining them within tolerable service limits for the foundations. The preloads are sand fills, which was calculated originally on the expected service loads during operations, but incremented to that height to accelerate treatment time.

[1] Consorcio BICA - Millán Ingeniería, Proyecto TCM, esteban.mayorga.marin@gmail.com
[2] Consorcio BICA - Millán Ingeniería, Proyecto TCM, Costa Rica

The instrumentation used on site were surcharge plates and piezometers, placed on the initial filling of the land reclamation area at the work fill level of the port, and prior to the preload to measure the consolidation of the deeper clayey soils. For settlement control, the project was divided into 18 work areas, which were preloaded in a sequence using the same material, but moved from one area to another. In consequence, the treatment time at each area was critical for the overall construction schedule of the project.

Settlement auscultation plates were placed on the initial filling of the port area without preloads, to then survey elevations in the same plate to obtain the settlements produced by the preloads. The information was taken in weekly measurements. The measured settlements were used as a basis for model calibrations and predictions.

So, the main objectives of the settlement data analysis were: 1) to verify the settlements monitored on the settlement plates and 2) to predict the remaining holding times necessary to achieve the target consolidation. Based on the information from the contractor, the designer and inspector of the project decides when was adequate to withdraw the surcharges at each area, for efficiency of the soil improvement and release of the areas for other activities on the construction schedule.

Consolidation was revised through the Observational Procedure of Settlement Prediction Asaoka (1978)[1], which is a method based on the plates settlement measurements, to calculate future settlement.

As part of the geotechnical verifications, a computational analysis of settlements for the final scenario was performed in a critical area of the project, simulating the scenario after the construction of the project, and compared with analytical method of Asaoka. This was a 3D model based on finite differences. The model includes stratigraphy and parameters for each layer, taken from the soil investigations, and representative boundary conditions.

Also, CPT tests were executed before and after, to verify with measures not only the consolidation degree of the soils, but also the resistance parameters for slope stability in the slope embankment below the pier. This will not be discussed in this paper.

The analysis evaluated if the required performance settlement would be achieved, according with the technical project specifications. The relationship between instrumentation and diagnostics is described as consecutive processes that are intrinsically related for effectiveness. According to that, for these authors is necessary to keep the instrumentation points (plates points and survey points) permanently to verify the predictions during the operation phase and not just in the construction stages.

1. DESCRIPTIONS

For this project several geotechnical investigations were performed. The general stratigraphic profile that is recognized in the wharf area is as follows: a) silty sand unit, b) alternating layers of sand and clay in the first half of the wharf, intercalating with a fat clay layer, and c) a mudstone unit.

The process of soil improvements includes Deep Soil Mixing (DSM), stone columns and (Vertical drains) PVDs, this last in an area of 36 hectares, from the level +2 m. The drains were installed in a triangular configuration. Also, it was necessary to use preloads to consolidate the clayey soils at the bottom of the reclamation area. The material used was mainly to sand dredging, in two ways:

a. Hydraulically (direct output of discharge pipe), in areas surrounded by previously constructed levees of the same material. This was done in the first treated areas.
b. In all the other areas, through earthmoving machinery, i.e. excavators and trucks, mobilizing the material from one area to another. This method was used mainly when the treatment was completed in one area and it was required to release the surcharge and translate it to another area.

2. INSTRUMENTATION AND DATA ADQUISITION SPECIFICATIONS

2.1 Settlement plates

To track preloads two monitoring systems were applied: a) an analysis of sampled data of several settlement plates around the reclamation and b) soil resistance by testing piezocone (CPT) results. This paper deals with the first revision.

The settlement plates were installed when filling in the reclamation area was above water sea level, and about +2 m from low water level (LWL). The measurements include georeferenced position, and the upper topographic level of the settlement plate and of the general ground level around the control points.

Finally, all data was analyzed and interpreted using analytical methods like Asaoka and computational models. Based on that, it was established if the residual settlement required by design has been reached, before authorizing the release of the area. This implies the unloading of the preload, operation that now had been already carried out in all the areas of improvement.

2.2 Settlements measurements specifications

The settlements thresholds established in the technical specifications of the project for the acceptance of preloads through the criterion of settlements for the yard area:

a. A maximum total settlement after 2 years of completion of the work
b. A maximum total settlement after 20 years of completion.

The procedure is to install the settlements plates and do an initial survey of on the initial filling without preload. When the preload is placed, measuring of the produced settlements are initiated. The instrumentation was placed from the start of monitoring in yard areas prefilled 1-6, 9-19 according to the numbering previously established (see *Figure 2*).

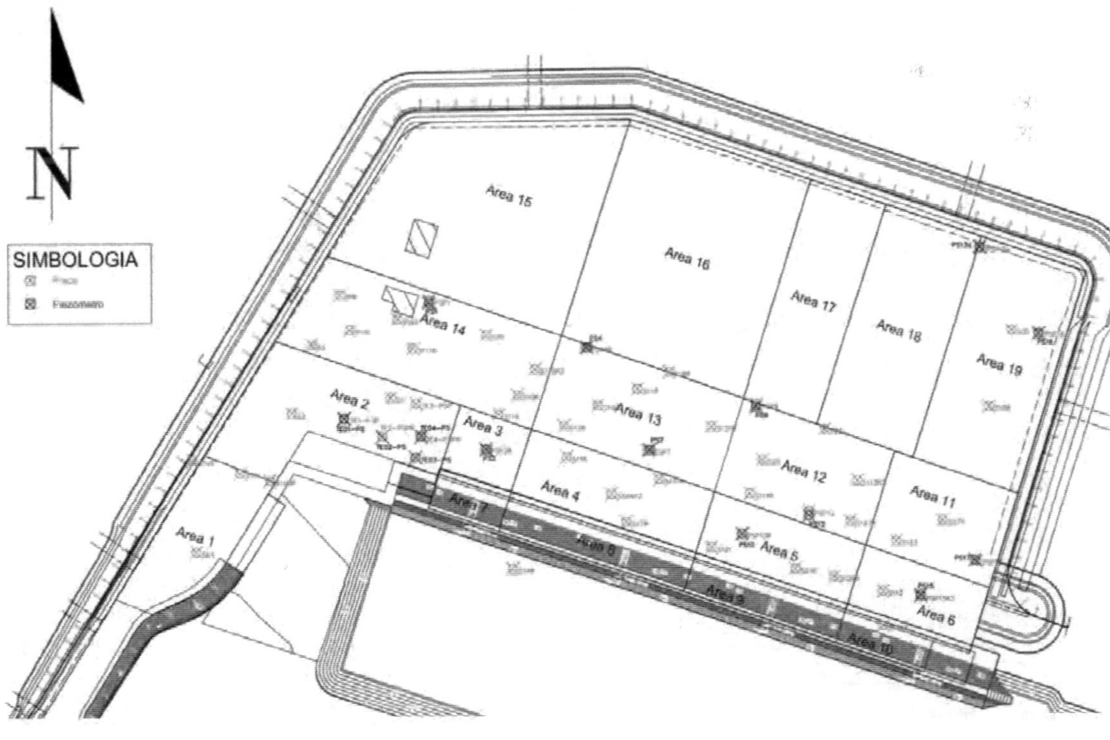

Figure 2: Sub areas preloads.

Measurements were made usually on weekly basis, i.e. sampling every 7 days, which is the minimum frequency recommended by the Asaoka's method. There were some weeks without measurement. This could be for several reasons, being the principal that there is not access, for example if the hydraulic fill is nor stable or for bad weather. In such cases, before applying the Azaoka method, data should be normalized, which means interpolation of the measurements so that the data is presented as settlement value every 7 days.

3. ANALYSIS AND PROGNOSIS OF SOIL SETTLEMENTS

3.1 Settlement analysis by Asaoka's method

It was performed several analytical calculations based on data acquired from settlement plates of the yard area. For this, as explained, sampling rate of seven days was required for applying the Asaoka's method for predicting the future settlements. Therefore, in some cases, measurements were interpolated between them to have a consistent methodology. The procedure is well known by the engineering community and it's explained widely by Asaoka (1978). Based on the analysis, the following observations was reached:

a. The maximum settlement criteria or thresholds requested in the technical specifications of the project were met for the soils corresponding to the settlement plates located in the control areas 1-6, 9-19.

b. In areas 17-19 the service settlement was reached prematurely, as a consolidation process took place over time thanks to the initial 2 and 4 m preload columns for long term, while the other areas were treated.

c. There are other specific cases that required other analyzes, derived from the analysis of future settlements that are not considered in this article as

    i. The treatment of the strip of the temporary breakwater for the construction of the patio included through the areas 13, 15- 17 and

    ii. The boundaries between the areas 6 of the yard and 10 of the wharf where the slope between the wharf and the yard is formed, which depends on the resistance to soil cutting and seismic analysis.

3.3 Computational settlement analysis for the final construction stage of areas 6 and 10

As part of the geotechnical validations it was carried out a computational analysis of settlements for the final stage of the project, after construction of the project. This to validate the analytical calculus memory realized with the Asaoka's method. The mathematical model that the software use is based on the finite differences method. The model includes all border conditions, such as stratigraphy and the different parameters of the different soil strata.

The parameters in the case of clays depend on the consolidation. When dealing with granular soils (gravel, sands) the elastic modulus of the soil is used, as the main parameter for the soil modeling. Also, this model included the improvements with DSM, stone columns, the steel piles, the concrete slab over the piles and finally the service load. In Figure 3 it´s shown a prognosis of a maximum settlement in 20 years, which is within the required specifications.

In Figures 4-6, show graphically the results obtained in the model. The settlements by consolidation reveal that there is very little permeability of clays, which makes the settlements reflected in the surface are considerably smaller than the settlements produced by the service load. This is, the graphs show a rapid consolidation in the time due mainly to the elastic modules of the improvements (DSM, etc.). The above validates the analytical model of Asaoka used so far for the analysis of the soil settlements, since it complies with the maximum thresholds of the settlements on the wharf and in the yard.

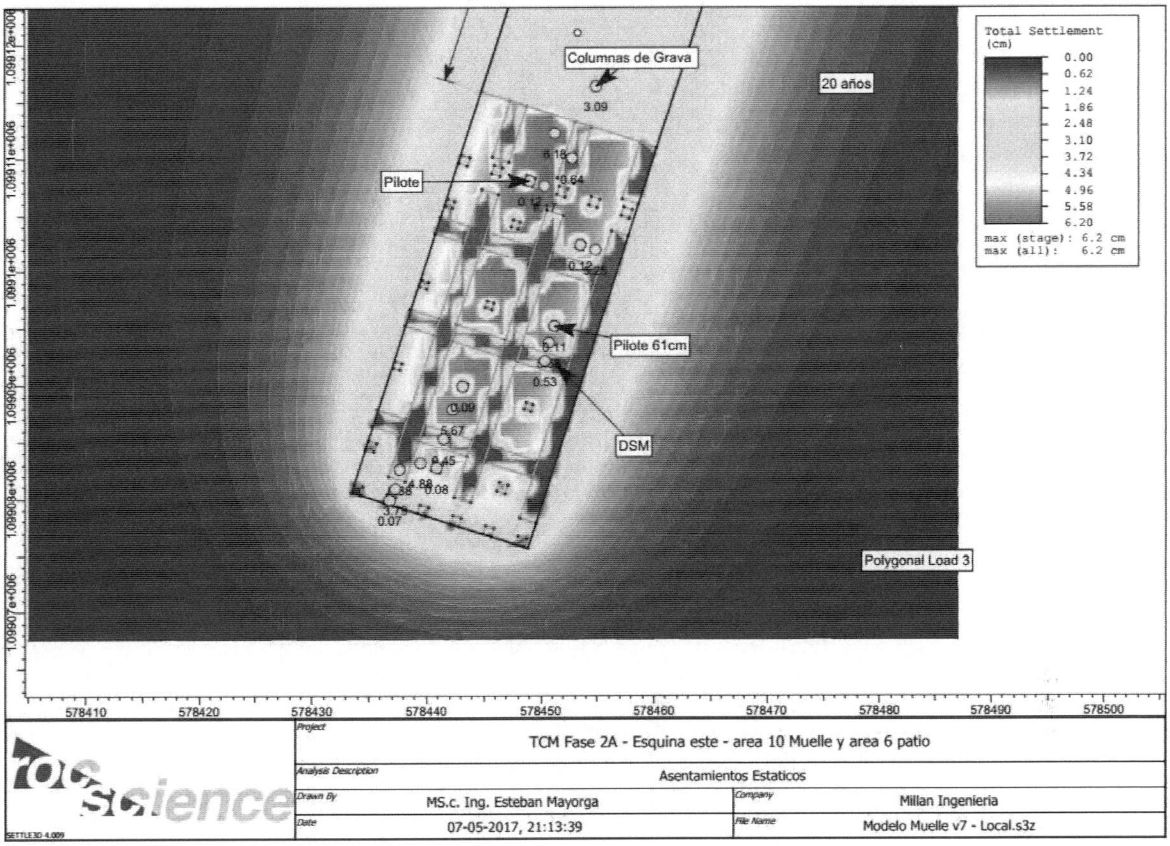

Figure 3: Computational model results for settlements predictions to 20 years in areas 6 (yard) and 10 (wharf).

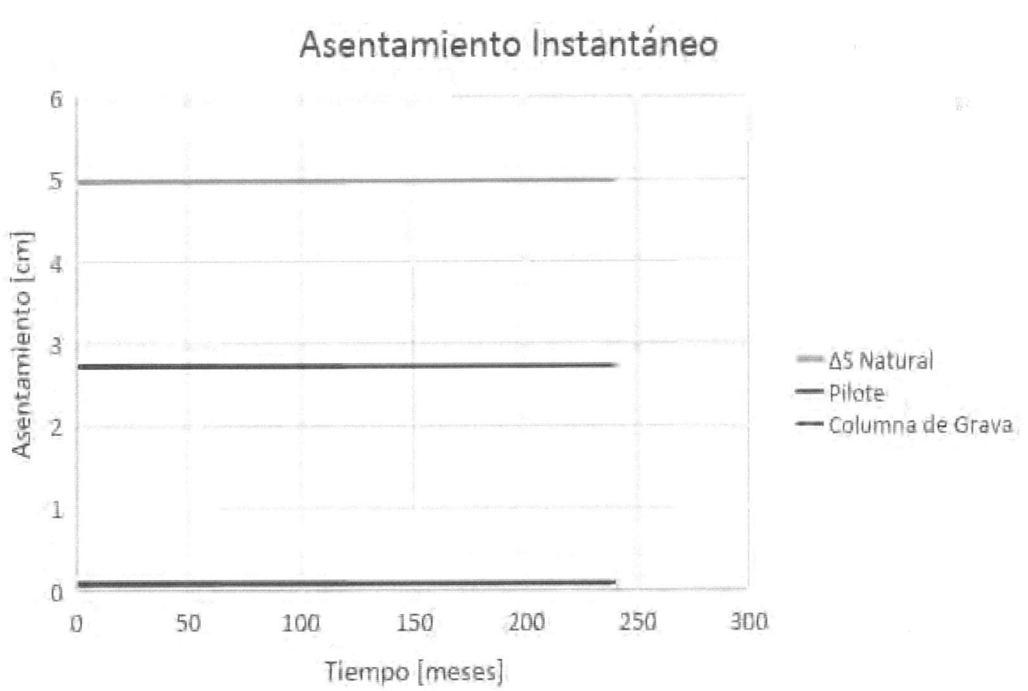

Figure 4 Instantaneous settlements due to service loading in areas 6 and 10.

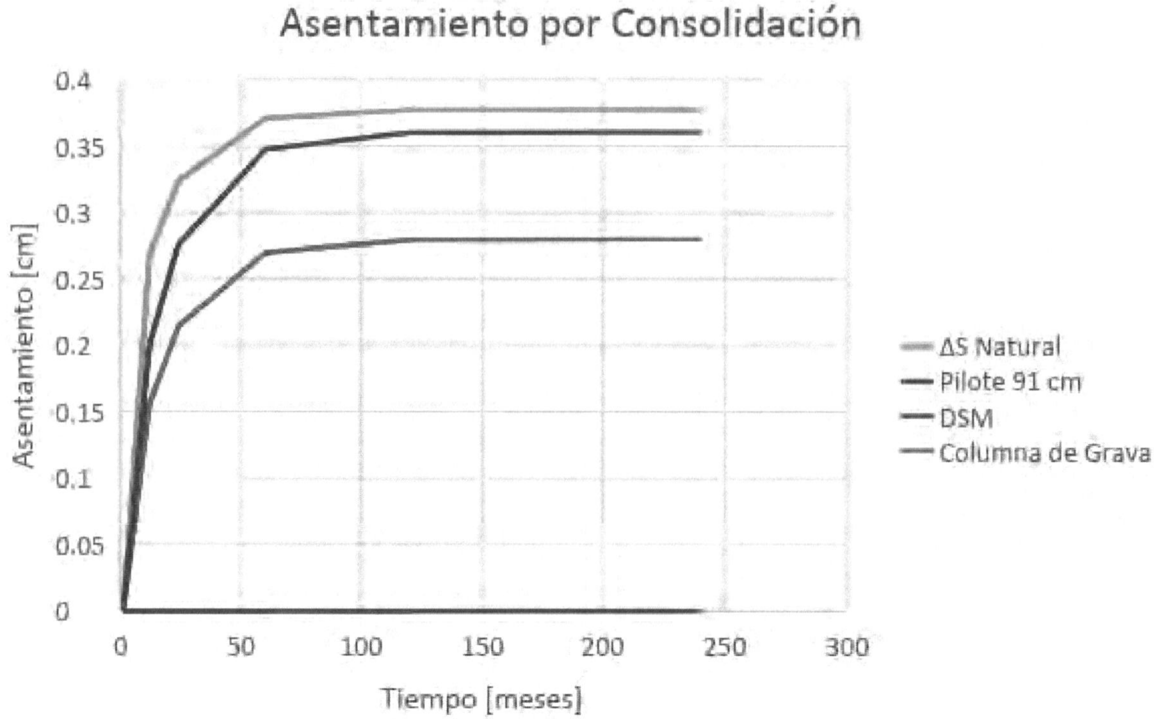

Figure 5 Settlements for consolidation over time in areas 6 and 10

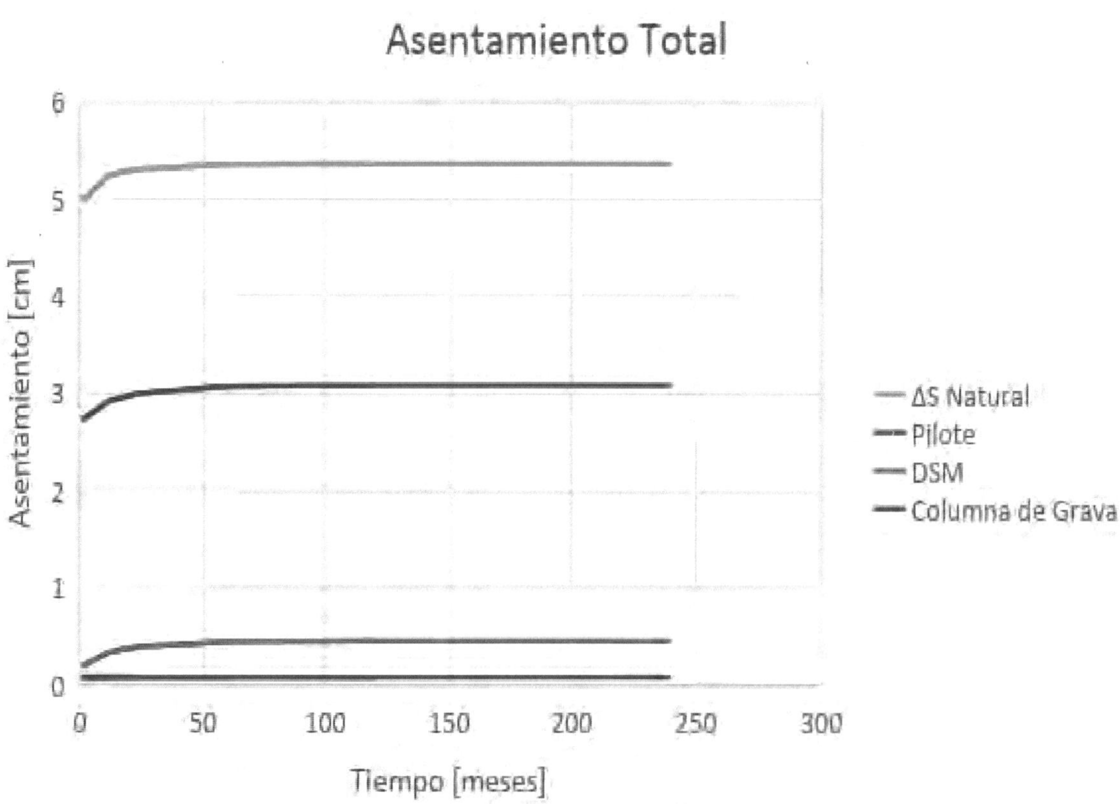

Figure 6. Total settlements (instantaneous + consolidation) in areas 6 and 10.

4. CONCLUSIONS

The settlements prognosis trough analytical and computational models, with boundary conditions, based on real data taken in the field is necessary for establish the different stages of the project. It improved the construction processes, to have the best decision-making, saving time and money.

In this project, as normal thinking, a plan was prepared for the instrumentation of the yard area in order to monitor and verify that the behavior of the settlements is within the thresholds defined in the technical specifications. Also, helps to verify if the model estimates remain realistic, or identify if additional modifications are required. It is necessary to maintain the instrumentation over the years, as well as the topographic records, as part of the maintenance and operational stages.

Follow up during the operation phase is important as a verification that the settlement threshold is really accomplished, and even in the case that they greater than expected, to apply the countermeasures for avoiding damages, principally of structures and pavements.

There should be control of the topographic points outside the project used for the measurements, as well as the guarantee the protection of these points for future revisions of settlements. This is because the settlements in the areas are significant, which modifies the levels and possibly positions of the control points.

5. References

[1] Asaoka, A. (1978). Observational Procedure of Settlement Prediction. Soils and Foundations Vol. 18, No.4.

PLANNING, DESIGN AND SUCCESSFULLY EXECUTING 4M M3 OF DREDGING AND DRY EXCAVATION TO EXPAND PSA PANAMA'S CONTAINER TERMINAL

By

David Taylor, MICE CEng[1]; Jan Neckebroeck[2] and Dominic de Prins[3]

ABSTRACT

Of all the obstacles in the construction of the Panama Canal over a 100 years ago, it was the complex, varied geology that possibly provided the biggest challenge, being the primary cause for the failure of the French Canal Company to construct the channel. When planning a new construction project involving removal of more than 4m m3 of material for expansion of PSA's container terminal at the Pacific Entrance of the Canal, obtaining information and how to structure the project was of prime importance.

In this paper the authors will explain the development of the project through the various stages, from concept design to successful execution, showing how knowledge and understanding of the site developed and how this changing information was used in the design, tendering and wording of the construction contract.

The solution adopted by the contractor, using a combination of dry excavation, together with backhoe, trailer and cutter-suction dredgers, was developed based on detailed experience gained from working in similar conditions nearby in Panama.

1.0 DESCRIPTION OF THE TERMINAL

1.1 Overview

PSA Panama International Terminal (PPIT) is located at the Pacific entrance to the Panama Canal on the west bank (the opposite side to Panama City). Figure 1 shows the location. The terminal operates under a concession with the Panamanian government, via the Maritime Authority (AMP). The initial concession was granted in 2008.

Planning, design, and construction of the terminal took place over a three-year period with the first cargo handled in 2011. The design was based on a single Panamax berth of 330-meter length, three quay cranes, and an operating area of 14 hectares, including container yard, gate, maintenance workshop, and administrative and customs buildings.

Figure 2 shows the operating Phase 1 terminal, with a large container vessel on the berth and well-utilized container storage yard. The administration building, gate complex, and equipment workshop are in the right-hand corner and clearance has started for the Phase 2 construction area.

[1] Head of Civil Engineering Latin America, PSA, david.taylor@globalpsa.com
[2] Regional Manager, Jan de Nul, jan.neckebroeck@jandenul.com
[3] Project Manager, Jan de Nul, dominic.deprins@jandenul.com

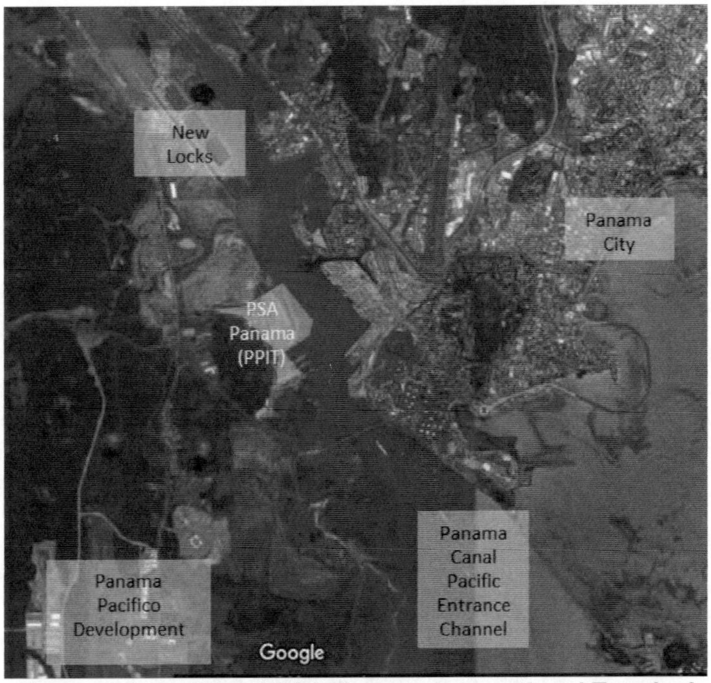

Figure 1: Location of PSA Panama International Terminal

Figure 2: Completed PPIT Phase 1 Terminal

1.2 Phase 1 Construction

Construction of the quay deck was carried out partially in the dry, with the rock slope at the rear of the quay wall being constructed in a bunded area. Figure 3 shows the initial excavation of the slope at the rear of the quay deck, together with a later photo showing the ongoing quay deck construction, including piling operations, placing of the grillage of concrete beams to tie the piles together. Figure 4 shows further construction photos during the piling and quay deck construction.

The construction was awarded to a JV of Colombian Contractors as a single contract, including dredging, quay, yard, electrical, utilities and buildings.

Figure 3: Phase 1 Quay Deck Construction

Figure 4: Phase 1 Quay Deck Construction

During the Phase 1 quay deck construction and dredging, an unforeseen outcrop of basalt was encountered, which caused significant delays and costs to the overall project, as well as altering the construction and dredging methodologies.

2.0 EXPANSION OF THE TERMINAL

In the period from the planning and negotiation of the concession in the mid-2000s, there have been many significant changes in the shipping world. There has been a dramatic increase in the size of container vessels (see Figure 5), the opening of the expanded Panama Canal, and major consolidation of container shipping lines and alliances (see Figure 6).

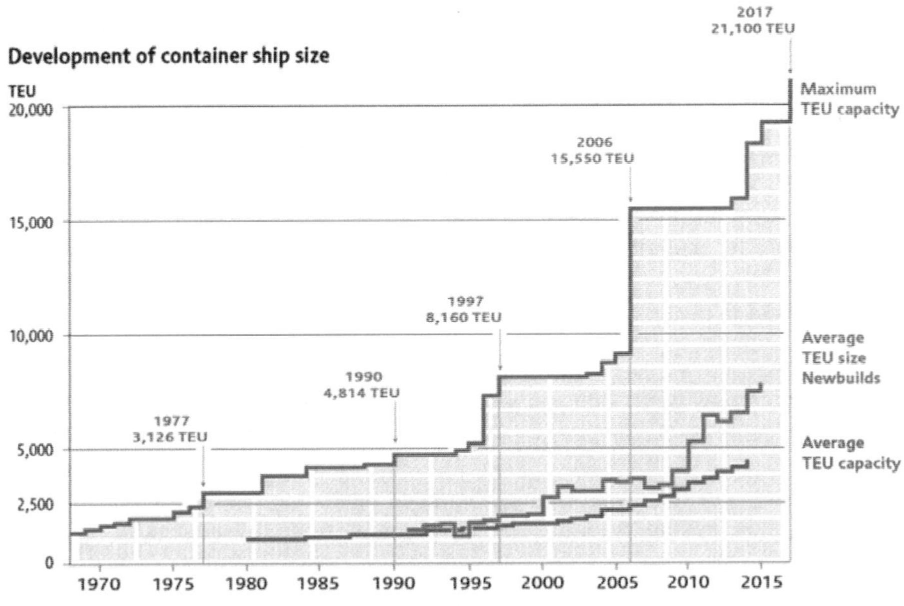

Source: The International Transport Forum

Figure 5: Evolution of Container Vessels Sizes

The dimensions of the locks in the original Panama Canal restricted vessel dimensions such that the maximum capacity of container vessels was approximately 3,500 TEUs (twenty-foot equivalent units – the standard measure of container volumes and capacities, with 1 TEU being a single ISO 20-foot-long container).

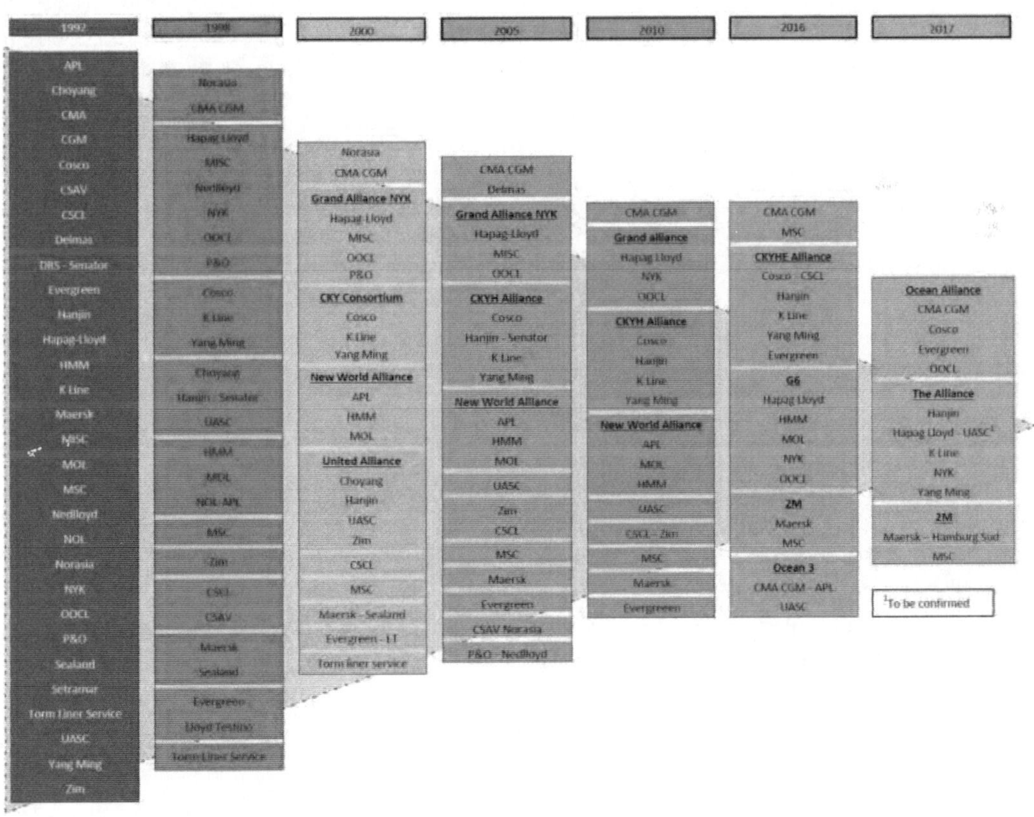

Figure 6: Consolidation of Container Shipping Lines and Alliances

Source – Ricardo J. Sanchez (UN-ECLAC)

The enlargement of the canal, which was completed in 2016, now enables vessels of up to 13,500 TEUs to transit the canal, greatly changing the economics of container vessel operations. In combination with this, there has been a huge increase in ship sizes, and the quantity of these large ships, as shipping lines have sought to reduce the shipping costs per container. The injection of capital needed for these new build campaigns and the need for revenues and capacity has resulted in dramatic consolidation of individual shipping lines and shipping line alliances to the point where there are now only three global alliances. Figure 6 illustrates this rapid consolidation starting from the mid 1990's to the present day.

3.0 DESIGN AND PLANNING

3.1 Preliminary Design and planning

The first stage in planning the expansion was to carry out vessel navigation studies to determine the optimum alignment of the new berth with respect to the Panama Canal. This was carried out locally using the SIDMAR navigation simulation center in Panama and with Panama Canal pilots.

Once the berth alignment was established, then it was possible to design the terminal layout to suit the operational goals, including vessel sizes, terminal capacity, etc. thus enabling preliminary engineering design to be carried out and the scope of the site investigation works to be defined.

3.2 Site Investigation

3.2.2 Introduction

As noted in the introduction, the ground conditions in the area of the Canal are challenging due to previous volcanic activity that has left basalt intrusions and metamorphic zones, combined with more recent alluvial and sedimentary materials.

A common perception with regard to the construction of the original Canal was that the French didn't have the correct equipment, having not anticipated that there would be rock excavation. As might be expected, the reality was a great deal more complex. In places the igneous rocks are extremely hard (UCS strengths of 100mpa are not atypical) and very well cemented with 100% RQD in borehole recovery. However, there is also huge variation in rock types and strengths over very small distances – within a 50m space the rock can vary from 100mps to 5mpa, and RQD from 100% to 0%. In addition, the tertiary sedimentary rocks which are encountered along the route of the Canal are not stable and many landslide have occurred resulting in huge movements of material – the Cucaracha Slide is perhaps the most well know. A slide of more than 10M m3 of material from this slide closed the Canal for 9 months after opening in 1915, and regular movements have continued (landslides closed the Canal on 26 occasions between 1914 and 1986).

Given the above, considerable effort was put into carrying out a thorough data collection program, within the financial restraints of a project that was still in the feasibility stage.

3.2.3 Site Investigation Program

A very preliminary set of boreholes was conducted in 2012, but little useful data was able to be collected. Following definition of the berth alignment and preliminary terminal layout, a seismic survey was carried out along the line of the quay deck and the area to be dredged. The fact that the berth line was set back from the canal allowed this to be executed quickly and cost effectively as it was a land based campaign.

Figure 7 shows a longitudinal profile from the seismic survey, with the variation in hard rock (the magenta colors) being easily identifiable.

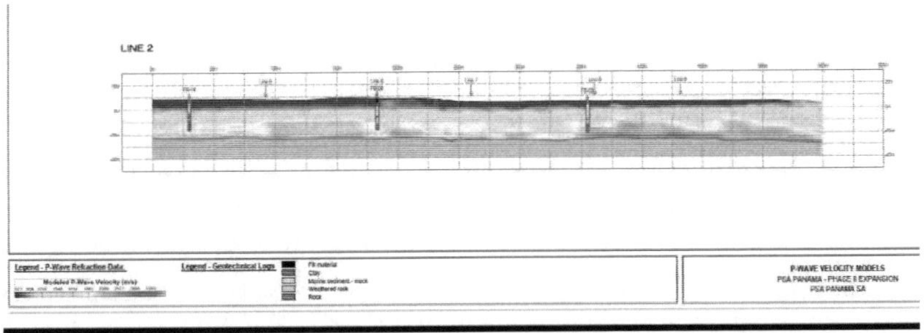

Figure 7: Longitudinal Profile from the Seismic Survey

Based on the results of the seismic survey, a scope of work was drawn up for a borehole and testing program. Following a competitive tender process, Fugro were awarded the contract, including around 20 boreholes (all land based) a CPT program to collect information on the soft sediments overlying the main site and full set of laboratory tests.

Figure 8 shows a borehole being drilled at the south end of the site and a photo of a core box of mainly basalt with almost a full core recovery indicating the material to have little or no faults of fissures.

Verification of the seismic and CPT's was carried out by conducting boreholes in the same location as some CPT's.

Figure 8: Borehole Drilling and Typical Core Box

Different laboratory tests were specified for the different materials – e.g. consolidation tests in the soft compressible materials, rock strength tests for the basalt and other hard materials.

3.2.3 Site Investigation Results

As has been indicated in the text, the results showed a changeable geology over the site, but with some basic patterns emerging.

Along the berth line there was an outcrop of hard basalt, with the top of it being somewhere around elevation +2m (relative to Chart Datum – MLWS) but that spread longitudinally around 200m and towards the dredged area by about 50m. At the south end of the site the rock elevation was around -18m and generally softer, being La Boca Formation (a sedimentary formation of weak mudstones, siltstones and limestones). At the north end there was a sheet of basalt at around -10m to -13m in a triangular shape with highly fractured limestone from the La Boca Formation.

The rock layers are overlain by Pacific Muck, a soft unconsolidated alluvial material containing a mixture of sands, silts and clays. This varied from 2 or 3m in places to over 10m in the north.

The Pacific Muck was in turn overlain by fill placed during the initial construction of the US Naval Base in the 1930's and dredged materials deposited since.

4.0 TENDERING PROCESS

4.1 Early Contractor Involvement

Given the known challenges with the ground conditions, an early involvement of potential contractors was sought to give as much time as possible for them to absorb the information about the site with the obvious aim of avoiding unrealistic tender prices. This was both in the sense of too high and too low.

A prequalification process was carried out, with the 2 main capabilities evaluated being hard rock dredging capabilities and experience working in these materials in Panama. It was important for PSA to have a dredging contractor who was used to working with the Panama Canal Authority and the rules and regulations pertaining to works in the Channel.

Boskalis, Dredging International, Jan de Nul and Van Oord prequalified. After signing a non-disclosure agreement with PSA, all available site was shared with the contractors and comment sought on the scope of the SI campaign being undertaken with Fugro. Contractors were also given the opportunity to witness the boreholes and laboratory tests, so being able to see first-hand the condition of the materials.

4.2 Tender Process

The tender documents were prepared using FIDIC Red Book (Engineer's Design) Conditions of Contract under the Laws of England and Wales. The design and tender documents were prepared by URS (now AECOM). Tendering was on a competitive basis, with price and program being the main focus.

An important decision in preparing the documents was to make the tender remeasurable. This was to avoid contractors including sums against the risk that the quantities of the different types of materials were different to those expected. Based on the available soil data (the Fugro SI was still ongoing) URS calculated quantities for 3 different stratum – Stratum 1: soft materials, Stratum 2: clay and weathered rock and Stratum 3: hard rock.

Another point of much discussion was the potential for of carrying out much of the dredging as dry excavation and the potential benefits it might bring with regard to certainty of materials encountered and control of costs. However, there was also a concern over the viability of the methodology and timing. Since it was not completed clear there would be a clear advantage of dry excavation over conventional methods, it was decided to allow either option to be proposed by the contractors.

3 tenders were received, with interview held with the best ranked 2 tenderers. Finally Jan de Nul were selected as preferred bidder and negotiations conducted with them. In joint venture with the maritime construction arm of Saipem (now Eiffage) Jan de Nul were also preferred bidder for the quay deck, based on an alternative design submitted by them and which was based on construction in the dry.

Through the negotiation process it was possible to change the contract to be executed on a lump sum basis. This was achieved through a combination of defining and apportioning risks to each party as appropriate and as acceptable to each party.

5.0 CONSTRUCTION PROCESS

5.1 Planned methodology

JDN opted to excavate a building pit first, in which then the quay could be built in dry conditions. In this way, the remaining land constituted a natural dike next to the Panama Canal, Once they quay was completed, the pit was flooded and the dike was removed by dredging. Such dry building pit offered a cost reduction in both the civil works and excavation works. It's also much easier to monitor the works as it's possible to just walk through the site for a visual inspection rather than when doing the works underwater. All excavated material, be it in dry or dredge, was transported over approximately 20km to an authorized off-shore disposal area.

The opted methodology divided the works in three main construction activities – Activity 1: pre-treatment of hard material in both dry excavation and dredging area, Activity 2: dry excavation and off-shore disposal of material but leaving a retention bund for the Panama Canal and Activity 3: Dredging and off-shore disposal of the remaining dike up to the design level of -16.3m MLWS. In between the dry excavation works and the dredging works, once the main piling works and corresponding excavation and rockworks in between the piles were finished, the flooding of the pit was to be carefully designed and planned.

5.2 Drilling and blasting program

Based on the site investigation results. pre-treatment of the hard material was required both in the foreseen dry excavation area with drilling and blasting of La Boca formation and basalt, and in the foreseen dredging area with drilling and blasting of only basalt as La Boca formation was expected to be dredgeable with the foreseen equipment.

Normal practice would have been to first remove the soft overburden before commencing drilling and blasting operations, reducing the amount of linear meters to be drilled. However, this was not possible for several reasons – the first reason: the overall construction schedule to meet the Contractual Milestones required drilling and blasting program to commence as soon as possible. Dry excavation depended on the installation of a temporary loading facility for loading into barges. If drilling and blasting were only to start after removal of soft overburden, this activity would become critical and increase overall execution time. The second reason was the interface with the quay piling works: a minimum distance of 100m was established between any blasting activity and the closest (finished or unfinished) piles to avoid damage. The drilling and blasting program was thus pushed forward to always guarantee this distance. The third reason and last was the design of the temporary retaining bund which had to remain intact; overburden could not be removed in this area anyhow.

During drilling works, an unforeseen basalt outcrop was identified outside the foreseen area and within the footprint of the temporary dike. Pre-treatment of this basalt area was handled separately as early blasting could lead to stability and permeability issues with the dike. The drilling of this area was done in the last weeks in dry on terraces before flooding of the excavation pit; blasting was done after partial flooding to avoid a sudden collapse of the dike due to blasting vibrations.

5.3 Excavation of the pit

All excavated material, be it in dry or dredged, was to be transported to and disposed at an authorized off-shore disposal area at approximately 20km from the site. In order to do so, a temporary loading facility was designed and constructed, consisting of a flat top barge fitted with four mooring pile collards and hinges for connecting loading ramps (see Figure 9). The collards ensured proper securing of the pontoon to four temporary concrete mooring piles that were casted up to rock layer. This floating jetty was a solid solution for coping with the high tidal differences at the Pacific side (5m) and the large layer of Pacific Muck that made other alternatives less favorable (e.g. sheet piles). The floating jetty was connected by means of three loading ramps to the shore and equipped with fenders on sea side for mooring a self-propelled split-hopper barge with a hopper volume of 3.700m^3. Two large excavators in a fixed elevated position would load the material that was brought to the jetty by means of articulated dumpers into the hopper barge.

Figure 9: Temporary loading facility for off-shore disposal

Each row of piles required a working platform at a different elevation and so the excavation of the pit (see Figure 10) was done in different phases and in close coordination with the piling works. Close to the first row of piles, dry excavation was done up to final design to avoid future dredging very close to the quay structure.

Figure 10: Dry excavation works

An access ramp was maintained at all times haul material to the temporary loading jetty. The biggest challenge was excavation and hauling of Pacific Muck in rainy weather conditions. As an optimization, when reaching below residual soil, weathered and blasted rock was recycled on site as general fill.

During the excavation works, the condition of the temporary dike was monitored periodically using inclinometers and drone footage to verify soil condition assumptions made when designing the dike and to detect anomalies such as erosion (piping) of a sand layer around 0m MLWS level.

The last activity before start of the dredging works was the flooding of the dry excavation pit once excavation, piling and rock works in the main section of the quay were finished. The flooding process was designed to be slow and controlled at the beginning and to faster once blasting of the unforeseen basalt outcrop underneath the dike was finished (see Figures 11 and 12)

Figure 11: Controlled start of flooding

Figure 12: Final flooding stage

5.4 Dredging to remove bund and complete depths

For the dredging of the dike and remaining material above final design level of -16.3m MLWS, a very complete dredging spread was mobilized consisting of a backhoe dredge with split-hopper barges, a cutter suction dredge and a two trailing suction hopper dredges. See Figure 14 for view of equipment deployed.

The backhoe dredge was used for removal of the soft upper layer of the dike (see Figure 13), for dredging in between the piles at the northern and the southern closing section of the dike and for dredging a trench in front of the first pile row wherever this was not done during dry excavation phase.

The cutter suction dredge was used for dredging of the La Boca formation and basalt layer to reach to design depth. Taking into account sailing distance and the expected variability of soil conditions, two large trailing suction hopper dredgers were used as barges with the cutter suction dredge for transport and disposal of the dredged material to the authorized off-shore disposal area. The loading of these hopper dredgers was done via floating line connected the hoppers bow connection. The large hopper capacity guarantied high efficiency of the cutter suction dredge independent of the type of material that was being dredged.

Apart from its use as barge with the cutter suction dredge, one of the two hoppers was used for general clean-up of the dredge area before its taking over and for maintenance dredging of Phase I berth pocket and basin.

Figure 13: Dredging of loose upper layer with backhoe dredge

Figure 14: Complete dredging spread at work

5.5 Accommodation of crane deliveries

As the start of dredging works was governed by the date of the flooding, which in turn was based around the critical milestone of the delivery of the first set of cranes. This was approximately one month after flooding and required a berth pocket for tide independent berthing and enough depth for the crane delivery vessel to reach this berth pocket. In order to reach this milestone, a section of 300m was selected in close coordination with the quay construction team and this section was dredged first with the backhoe dredge.

For the delivery itself and the commissioning of the cranes, dredging of the remaining scope was planned in such a way to minimize interference.

5.5 Variation in materials versus expected at time of tender

Although in terms of area less hard material was found than expected (which generated a saving on the drilling and blasting works), the hard material that was present turned out to be averagely harder than expected. This generated an extra cost for the dredging works mostly as a result of excessive consumption of soil wearing parts, requiring additional mobilization of spare parts and consumables and intensive repairs during and after the works. Thanks to the high efficiency of the dredging spread, JDN was able to maintain dredge productions within the scheduled execution time.

Additionally, an unforeseen outcrop of basalt was found under the central section of the temporary dike. This area was blasted after partial flooding of the excavation pit to ensure dredgeability.

5.6 Equipment solutions to accommodate variability of materials

The variability of materials that would be encountered was taken into account when selecting the equipment for this job. During dry excavation on one hand, the deployment of articulated dump trucks, the design of the loading facility for off-shore disposal and the type of barge guaranteed high production rates even in rainy conditions and with wet, soft material as the Pacific Muck.

For the pre-treatment of the rock, systems were foreseen for drilling with casing through soft overburden as well as drilling without casing straight from rock level.

On the other hand during dredging works, the selection of the powerful cutter suction dredger 'Niccolo Machiavelli' is with 7,000 kW of cutter power part of the world's most powerful cutter dredgers and was able to dredge the hard blasted and unblasted rock. The hopper barges made sure enough loading capacity was available and hence minimized downtime while changing barges or waiting for barges, not depending on the type of material encountered. On shore, a 24/7 welding workshop was deployed for continuous repairs to cutter heads after being worn or damaged by the hard material. Dredge planning was made in such a way that the backhoe dredge was working on-site until the end of the project, for the removal of some remaining big boulders that would have been tossed around by the cutter dredger.

6.0 COMPLETION

The execution of the dredging and excavation contract for PPIT was accomplished successfully and expeditiously. Final outturn price was less than 5% above the agreed contract amount. The only unforeseen event that occurred was a delay to the quay contract caused in connection with permitting for the concrete batch plant, which in turn impacted the start date of the dredging works. Otherwise the additional amounts were either due to increases in scope or other variations. Given what is known about the challenging soil conditions, this is a very significant achievement and a validation of the selected tender and contact methodology, as well as the execution of the works by the contractor.

Contractual Model for Dredging Projects to Avoid Disputes: Case Studies of the application of Performance Based Contracts in dredging projects around the world.

Prieto-Portar, L.[1], Prins, J.[2] and Prieto y Muñoz, L.[3]

ABSTRACT

The new application of **Performance Based Contract** (PBC) to dredging (also known as a "**draft guarantee contract**") transfers to the contractor all sedimentation risks over long periods of time and pays for all his services via monthly fixed amounts. The method is so successful that in some cases the PBC has lead to novel financing via tolls of the waterway (a **concession contract**).

Up to now, most dredging by port and waterway authorities has been paid from public funds through, (1) surveyed volumes and fixed price or, (2) a charter contract paid directly by the agency. These methods are known to harbor the risk of disputes, primarily arising from the technical documents when encountering adverse site conditions. PBCs are especially advantageous to both the Owner and the Contractor when: (1) there is a stable level of policy, (2) where sedimentation can be evaluated within reasonable assumptions of risk, (3) where all other risks are covered with reasonable contractual limits (for example, extreme weather conditions), and (4) there is a sufficient yearly volume to be dredged that allows for competitive bidding.

This paper provides supporting information of three successful applications of PBCs in Argentina and The Netherlands in large dredging programs. In all instances, disputes have been absent throughout these three contracts (involving over 20 years of continuous dredging). Additional data is presented on the technical and quality aspects that are most significant to the PBC to preserve the quality and successful completion of the projects.

Keywords: Performance Based Contract, adverse site conditions, dredging.

INTRODUCTION.

Performance Based Contracts are really a derivative form of Public Private Partnerships (PPP). This powerful financing method speeds up investments in intermodal infrastructure works and services. Although the PPP method has recently received wide coverage in the press and in technical magazines, it is now 355 years old!

In 1663, by Act of Parliament, a toll road was built via a PPP at Wadesmill in the Hertfordshire, United Kingdom. Throughout the 1800's, toll roads, bridges and tunnels were built in the USA using the PPP method, which was equally popular building railroads in France, the London subway, and even both major canals at Suez (1860) and Panamá (1880). The PPP method went out of fashion during the Great Depression (stemming from Wall Street's crash on 29 October 1929 and lasting through 1939). Following the Great Depression all infrastructure works became a state-owned monopoly until the resurgence of PPP in the 1980's when national budgets were allocated almost entirely on social programs and practically nothing on infrastructure.

[1] Chief Port and Geotechnical Engineer, Piedroba Consulting Group, Miami, FL / dr.prieto@piedroba.com;
[2] Vice-President, Piedroba Consulting Group, Miami, Florida / prins@piedroba.com;
[3] President, Piedroba Consulting Group, Miami, Florida / prieto@piedroba.com.

This paper discusses the use of the PBC method via data from three successful PBC dredging projects in Argentina and The Netherlands.

CASE #1: Bahia Blanca, Argentina.

Figure 1: General location map of the two port areas using the PBC method in Argentina: (1) at Bahía Blanca and (2) from the Atlantic Ocean - Río de la Plata - Río Paraná to Santa Fé (red line).

The 97 kilometer length of the Bahia Blanca estuary has 45 major cargo terminals. In 2016, these terminals handled 27.400.000 tons of cargo from 1,051 vessels.

Port operations are greatly affected by very high tides, from 4.5 m (15 feet) at the port down to 2.5 m (8 feet) at the channel entrance in the Atlantic Ocean. The minimum draft is 13.7 m (45 feet), which must take into account these very high tides. The port's management identified a "window" of operations in order to reduce the annual cost of dredging, in combination with the use of Performance Based Contracts (PBC).

Before 1980, the national Argentinean government operated all of the ports in Argentina. From 1980 to 1999, port operations were transferred to their respective provincial governments through new local public agencies. These agencies let out dredging contracts based on pre-fixed volumes. These pre-fixed volumes led to problems maintaining their minimum drafts, and there was no guarantee that a vessel could traverse any particular channel without grounding.

For the above reasons, since 1999 the Port of Bahia Blanca adopted the use of Performance Based Contracts (PBC). That decision was based on:

(1) The many year of experience of Bahia Blanca's port personnel;
(2) The meticulous gathering of massive hydrodynamic data of the port's conditions throughout the year during 20 years, including detailed knowledge of the types of sediments;
(3) Preparation of statistical analyses of that data to predict future volumes of sediments.

The results of using PBC since 1999 has been that the minimum draft of 13.7 m (45 feet) has been maintained during 24 hours a day, 365 days of the year. That draft was paid solely from the tariff fees levied upon the port users.

These PBCs were granted in 5-year increments. Their implementation was based on three phases:

- Phase One: In **Year-1**, the channel was fully **surveyed** (bathymetry) to establish the pre-dredging conditions. The contractor **mobilized equipment and personnel**, pulled all the required **permits**, established field and office operations, etc.

- Phase Two: **Capital dredging also in Year-1 to attain the minimum draft and channel width**, with an initially pre-estimated volume of 1 million m^3 based on hydrodynamic studies. This phase also included placing all navigational aids. After this phase was concluded and the actual volume dredged was ascertained to reach the minimum draft of 13.7 m (45 feet), then the payment was adjusted accordingly.

- Phase Three: **Years 2 through 5 involved maintenance dredging, improving the navigational aids and performing correctional studies to adjust the channel.**

The contract consisted of 60 equal monthly payments, plus the initial adjusted capital dredging.

The consequences of this PBC was that the contractor assumed all the sedimentation risks over longer periods of time and was paid for his services with monthly fixed amounts.

As the contract evolved, the PBC converted to a Concession Contract, where, in addition to sedimentation and draft risks, the contractor's monthly payments were made entirely via user tolls.

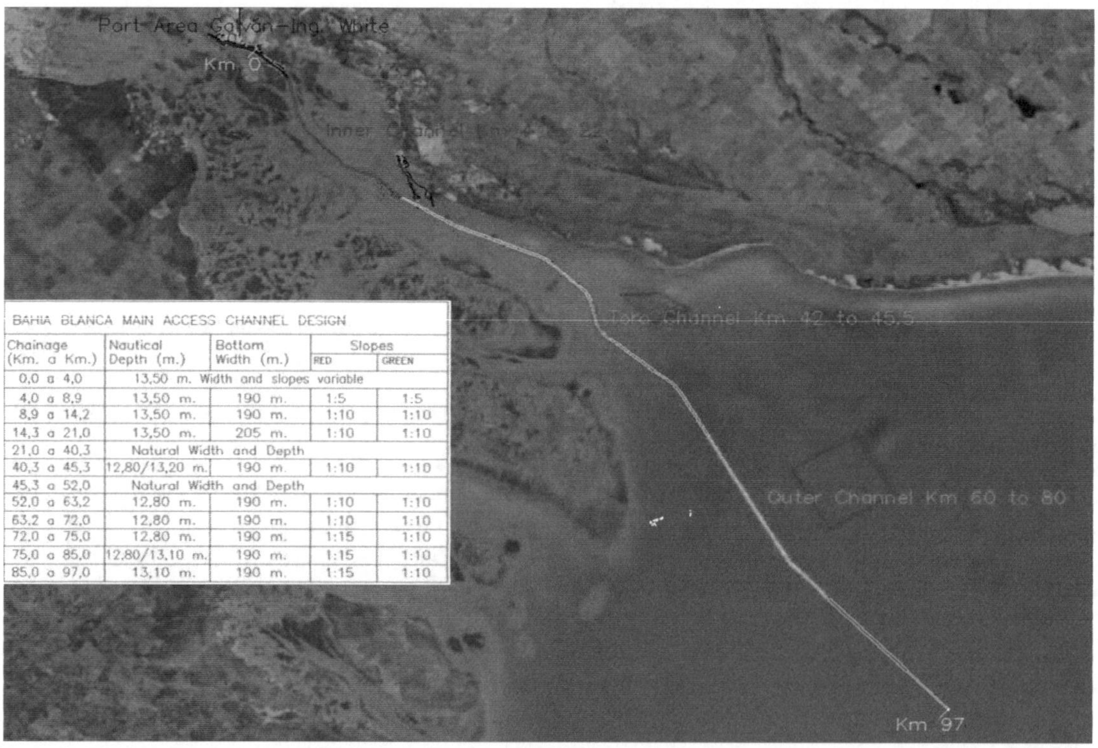

Chainage (Km. a Km.)	Nautical Depth (m.)	Bottom Width (m.)	Slopes	
			RED	GREEN
0,0 a 4,0	13,50 m. Width and slopes variable			
4,0 a 8,9	13,50 m.	190 m.	1:5	1:5
8,9 a 14,2	13,50 m.	190 m.	1:10	1:10
14,3 a 21,0	13,50 m.	205 m.	1:10	1:10
21,0 a 40,3	Natural Width and Depth			
40,3 a 45,3	12,80/13,20 m.	190 m.	1:10	1:10
45,3 a 52,0	Natural Width and Depth			
52,0 a 63,2	12,80 m.	190 m.	1:10	1:10
63,2 a 72,0	12,80 m.	190 m.	1:10	1:10
72,0 a 75,0	12,80 m.	190 m.	1:15	1:10
75,0 a 85,0	12,80/13,10 m.	190 m.	1:15	1:10
85,0 a 97,0	13,10 m.	190 m.	1:15	1:10

Figure 2: The access channel for Bahia Blanca is 97 kilometers long with 45 major cargo terminals.

The latest 5-year PBC was signed on the 17th of February 2012 (renewed last year) with Royal Boskalis Westminster n.v., and notice to proceed (NTP) was issued on the 4th of April 2012. The PBC required dredging 3 million m^3 each year, which was equivalent to a cost of USD 72,000,000 per year. The cost of dredging was paid entirely by the tariffs levied on each ship that used the channel.

The PBC was made up of two parts: (1) the Technical Offer, and (2) the Financial Offer.

The Technical Offer required the bidder to fulfill a number of formal requirements and comply with the Technical Specifications. For example, there were two types of dredging; (1) the main channel was dredged with a trailing suction hopper dredge (TSHD) whereas the piers and wharfs were dredged with a water injection dredge (WID). When the contractor presenting the best Technical offer was identified, then the evaluation process passed to the Financial Offer.

The Payment Mechanism.

As an example, assume the project has an initial volume to be dredged of 2 million m^3 at a bid price of USD 10.50 per m^3, which has to be a competitive price included in the Financial Offer.

$$Monthly\ payment = \frac{(\frac{\text{USD } 10.5}{\text{m}^3})(2\text{x}10^6\ \text{m}^3)\text{+Annual maintenance costs}}{\text{60 months}}$$

$$Monthly\ payment = \frac{\text{(USD 21,000,000)+(4 years)(USD 4,500,000/year)}}{\text{60 months}}$$

$$Monthly\ payment = \frac{\text{USD 39,000,000}}{\text{60 months}} = \text{USD 650,000/month}$$

Figure 3: The dredging of the main channel for Bahia Blanca's 97 kilometer estuary was done by the TSHD "Shoreway" from Royal Boskalis Westminster n.v. (hopper capacity 5,600 m^3).

Figure 4: Dredging of the piers, wharfs and quays was done by a water injection dredge (WID).

In conclusion, the Port operators attribute the success of the PBC method at Bahia Blanca to the following reasons:

(1)The PBC method guaranteed the safe navigation of all their client ships with an insured minimum draft in the entire length of the 97 kilometer channel, during the entire year without exclusion times;

(2) A major factor that insured the method's success was having accumulated data during several decades of good and bad experiences of every aspect of the port terminals and the entire estuary.

(3) The accumulated data of the last 20 years of port operations was then statistically analyzed and the results of the analyses predicted the rates and types of sedimentation, which were shared with the bidders for their use.

(4) The PBC experience has shown that much lower bid prices are obtained when the contracts are at least five (5) years long.

(5) The PBC was further enhanced if the port authority guaranteed monthly payments in-full.

(6) It is important to maintain a cordial relationship with the contractor. For example, the port personnel did not rush to impose penalties for infractions. It was better to establish simple legal rules to resolve all contractual problems in an atmosphere of mutual respect and partnership.

In parallel to the Bahia Blanca experience, Royal Boskalis also held similar PBCs to dredge the Paraná and Uruguay rivers close to Diego García Island on the Río de la Plata, with equal success.

CASE #2: Atlantic Ocean, Río de la Plata, Río Paraná to Santa Fé, Argentina.

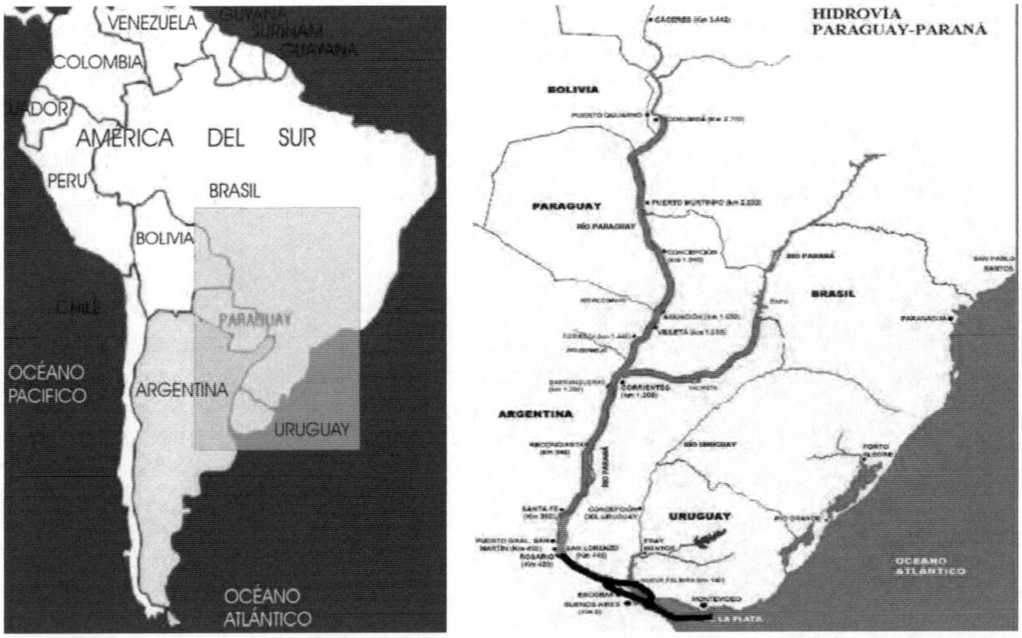

Figure 5: (Left) Location of project's influence area in South America; (Right) Location map covering the PBC from the Atlantic Ocean, up the Río de la Plata, Río Paraná to the port of Santa Fé.

An even older PBC success (by four years) took place on the Río Paraná, managed by the contractor Hidrovía Waterway Concession (led by the Belgian dredger Jan De Nul) between 1995 to 2013. That contract covered the much longer 800 kilometer channel: from the Atlantic Ocean to Río de la Plata to Río Paraná up to the port of Santa Fé. This navigational channel carried over 80% of Argentina's exports, and included not only ocean-going traffic but also heavy local river barges..

In most respects, the PBC had the same elements as the Bahia Blanca PBC. The contractor assumed the risks of sedimentation and the consequent increase in channel traffic.

The first and second phases were similar to Bahia Blanca's, with the capital dredging of 22 million m^3 performed from 1995 to 1996 to deepen the draft of the 800 kilometer main channel from 9.8 m (32 feet) to 10.4 m (34 feet). In parallel, the entire length was covered with navigational aids (buoys, radio beacons, etc.). The PBC called for the maintenance of the minimum draft all 24 hour per day year-round without interruptions. A toll system was created to pay for the dredging and the navigational aids.

The contractor had operating three full-time trailing suction hopper dredges (TSHD) with hopper capacities ranging from 3,400 to 6,000 m^3 to maintain the channel year-round, but in peak sedimentation periods it brought in an additional large cutter suction dredge (CSD) to work heavily shoaled areas of the channel. The CSD also created sedimentation traps in the Río de la Plata estuary to capture peak sedimentation and guarantee the minimum draft at all times. Hidrovías also continuously maintained the buoys and electronic equipment along the entire length of the concession. Five (5) survey vessels continuously mapped the draft of the entire channel.

The tolls are paid by all commercial vessels based on their tonnage and length of channel navigated. These tolls are continuously revised to balance the costs versus the revenues to reach a net zero income. The costs for the typical user is about 1 USD/ton of grain exported or 16 USD/TEU.

CASE #3: The Rijkswaterstaad (RWS) of The Netherlands.

The Rijkswaterstaad (RWS) is the public executive agency of the Dutch Ministry of the Infrastructure and the Environment that manages all of the nation's territorial waters.

The Rijkswaterstaad dredges 30 million m^3 each year, with about half representing maintenance dredging (see Figure 6), whilst the remaining half is divided between capital dredging and beach nourishment. All dredging contracts are done by private contractors (the RWS has zero dredges) and all contracts are PBCs.

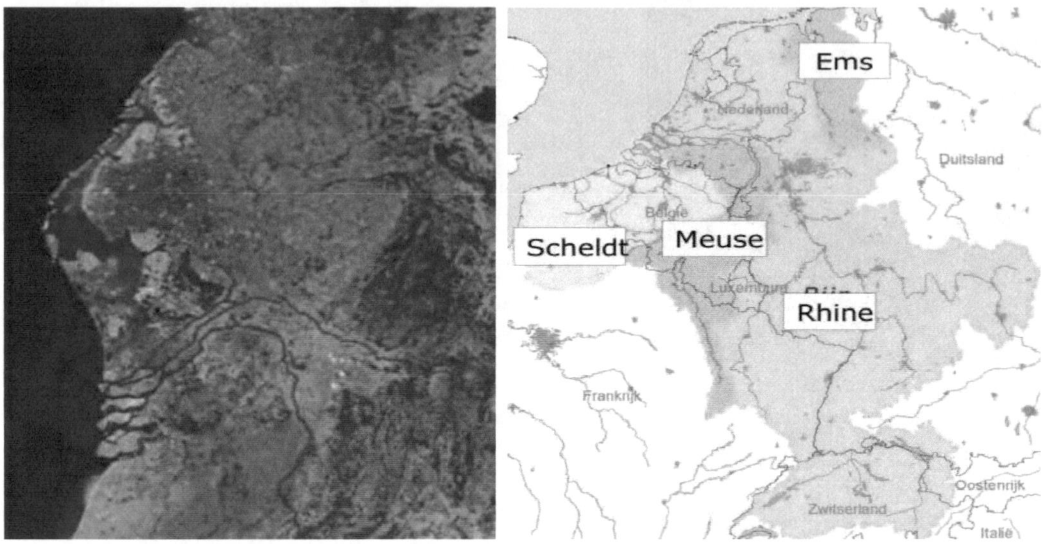

Figure 6: The Netherlands occupy the delta of four rivers with a population of 17.1 million (2016).

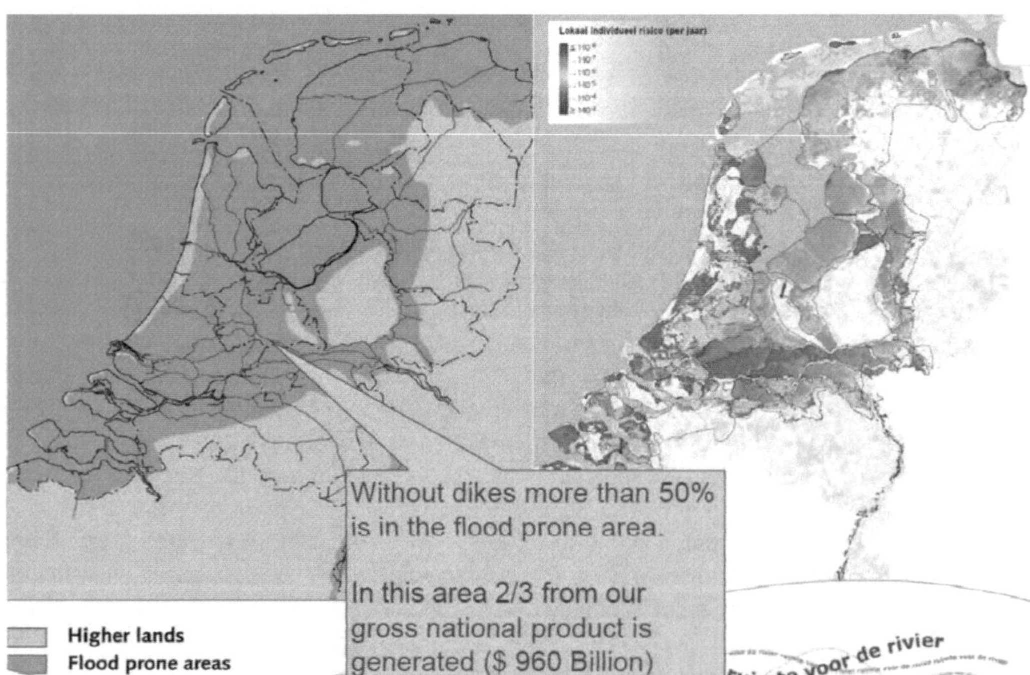

Figure 7: Without its dikes, over 50% of The Netherlands would be under water.

Like all PBCs world-wide, the RWS requires that their contractors assume all the risks of a dredging project. A typical organizational chart is shown in Figure 8, where most of the activities belong to the contractors who are thus responsible for all risks.

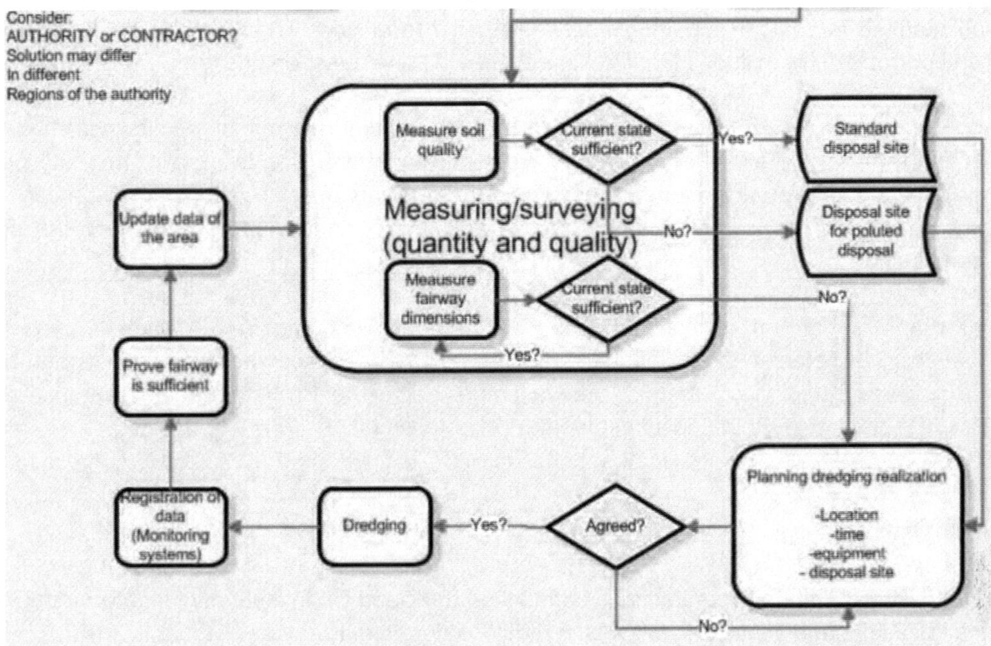

Figure 8: The RWS demands that all major decisions in dredging (yellow rectangles) must be taken by the contractor (the grey rhombs are decisions made by the RWS).

Figure 9: The RWS has stopped using traditional contracts and have moved to Performance Based contracts (PBC) where the contractor has most responsibilities. For example, their PBS requires the contractor to guarantee drafts in their canals and ports in exchange for maintenance contracts over longer periods of time.

The World Bank and Port reform.

The World Bank continues to be a major force in reforming the port sectors of emerging economies. Their **Service Port to Landlord Governance** model has found wide application, where the Landlord model has resulted in a clear separation between public and private responsibilities embedded in concession contracts. This includes the investments, tasks, revenues, and risks.

The World Bank provides decision support tools to undertake reforms of public institutions that provide, manage or regulate port services in developing countries. The Bank also helps to choose options for private sector participation and assists in the analysis of the relationships between public and private parties, suggests legislation, contracts and institutional charters to govern the private sector's participation.

The World Bank is currently implementing a PBC for the dredging in Bangladesh of a major navigational channel between Chittagong and the capital Dhaka. The small country is formed by the confluence of three huge rivers merging into a single delta: the Brahmaputra (Jamuna), Ganges (Padma) and Meghna rivers that discharge into the Bay of Bengal.

REFERENCES.

- Queiroz, C., "Public Private Partnerships as an Option to Speed Up Investments in Intermodal Infrastructure and Services", InterBaltic Final Conference, Klaipeda, Lithuania, 2007.
- Prieto-Portar, L., "Primer Curso de Dragado", manuscript, 2018.
- http://rru.worldbank.org/besnapshots.
- (On World Bank Port Reform Toolkit): http://web.worldbank.org/WBSITE/EXTERNAL/TOPICS/EXTTRANSPORT/EXTPRAL/0,,contentMDK:20517158~menuPK:1221870~pagePK:148956~piPK:216618~theSitePK:338594,00.html.
- (On Public Private Partnerships): http://rru.worldbank.org/Toolkits/Documents/Advisors/Full_Toolkit.pdf.
- (On Labor Issues in Infrastructure Reform): www.ppiaf.org/Reports/LaborToolkit/toolkit.html.
- (On World Bank's "Resource Guide for Performance-based Contracting."): http://www.worldbank.org/transport/roads/resource-guide/index.html.

WATERWAY ASSET MANAGEMENT WITH HOLISTIC DREDGING AND SEDIMENT MANAGEMENT ON THE DANUBE

by

Markus Hoffmann[1], Alexander Haberl[1], Thomas Hartl[2], Christoph Konzel[2], Markus Simoner[2] and Gerhard Klasz[3]

ABSTRACT

Recent research on traffic development and transport economics in the Danube corridor revealed the importance of improved navigation conditions on the waterway Danube. The developed Waterway Asset Management System (WAMS) of viadonau allows a real-time calculation of fairway availability, planning of dredging measures, sediment management and budget estimations for various target conditions. The paper provides an overview of the WAMS with focus on dredging and sediment management in this software tool reproducing the entire dredging process, starting from an automated analysis of critical sections, planning of dredging measures per drag and drop, automated cost estimations based on economy-of-scale cost functions and an overview on the status of all measures. Prior to the developed solution estimations of dredging volumes had been based on single-beam surveys and the profile method or a more accurate "manual" assessment of multi-beam data in ArcGIS. With these new capabilities an analysis based on accurate multi-beam riverbed surveys is feasible accounting for deviations in billing of conducted dredging measures. Further functionalities include an analysis of dredging impact duration based on sedimentation and erosion rates, enabling an assessment of the efficiency of a dredging measure in comparison to other possible solutions (e.g. fairway alignment, construction of groynes). Thus, a much faster and efficient planning, implementation and controlling of dredging measures in order to achieve higher fairway availability has been realized. Based on the concept for a systematic sediment management the paper also gives insight into the key elements, findings and functionalities of an advanced sediment management taking into account both economic and ecologic factors. By facilitating an analysis of sedimentation and erosion rates on short sections for a fixed time frame as well as the development over time including dredging and dumping measures the developed WAMS provides an overview on all conducted dredging measures and related dumping sites for a given time frame as well as a total balance on all erosion and sedimentation volumes for any given time frame and river stretch (sediment balance). Instead of lengthy analyses and studies the Module Sediment Management allows viadonau to constantly assess and adapt their approaches by optimizing both the selection and the timing of appropriate measures. In summary, the developed functionalities enable an efficient balancing of both the interests of environmental protection and inland navigation at the same time. The comprehensive analysis and documentation system is constantly being updated based on previous results, thus becoming more accurate with every year.

KEYWORDS: Inland waterway Danube, dredging management, sedimentation and erosion, balancing and compensation

---

[1] Vienna University of Technology – Institute of Transportation, Austria.
markus.hoffmann@tuwien.ac.at
[2] viadonau – Austrian Waterway Company, Austria
[3] KLASZ - Hydrology and Hydraulics Consultants, Austria

1. INTRODUCTION

1.1 Investing in transport infrastructure and IWW (OECD – Countries)

Transport infrastructures as important assets of national economies develop in very long cycles due to their service lives and costs whereas transport of goods (and passengers) are closely related to economic development. The connection of investments in transport infrastructure and economic development is well documented. Like all production functions the benefits of additional spending are decreasing with high levels of income and already developed transport systems. Nevertheless, the spending for high level transport infrastructure increases almost linear with income amounting to roughly 1% of total GDP. The over-view on high level transport infrastructure investments in Figure 1 shows, that the majority of investments goes into road (~60%) and rail (~30%). The remaining share (~10%) is divided between Air-ports, harbor structures and inland waterways. Despite a number of EU backed initiatives to shift goods transport from road to rail and waterway in-creasing spending and market shares for these modes are not to be expected under current circumstances (Hoffmann 2018).

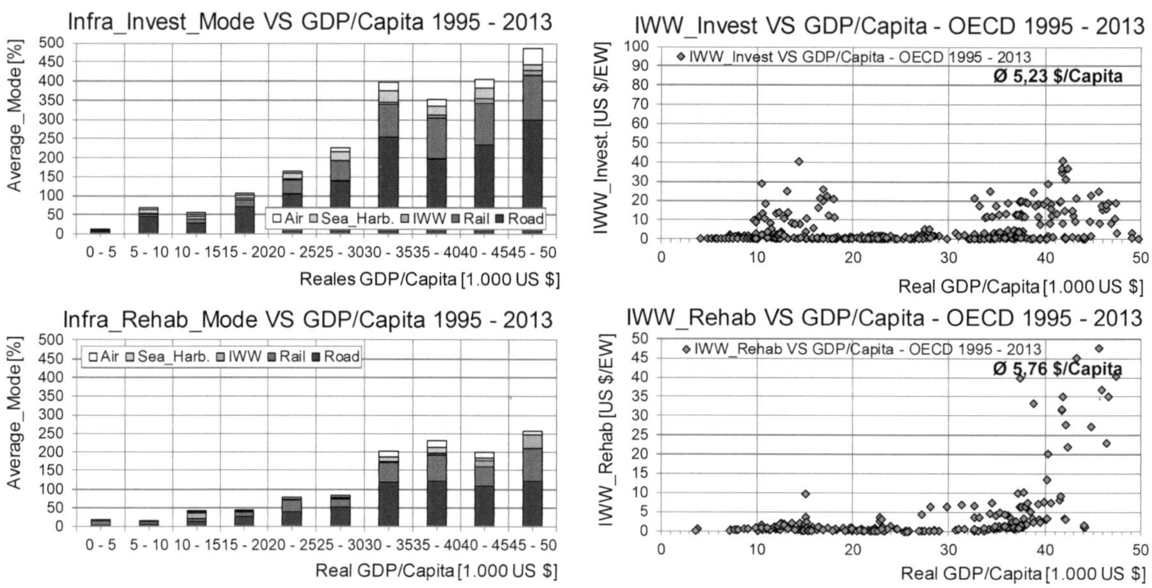

Figure 1: Total transport infrastructure investments and investments in inland waterways in OECD – Countries per capita VS Real GDP/Capita 1995-2013

1.2 Investing in IWW (Austria)

Managing the waterway Danube in Austria falls in the responsibility of the Austrian Waterway Company viadonau. The main tasks of viadonau are transport infrastructure and traffic management, flood control, real estate and other tasks (e.g. maintaining towpaths). In addition, viadonau spends a sufficient part of the budget for development and innovation (viadonau 2016, 2017). The budgeting shown in Figure 2 is variable depending on necessary additional expenditures (e.g. flood control, dredging 2012/13/14).

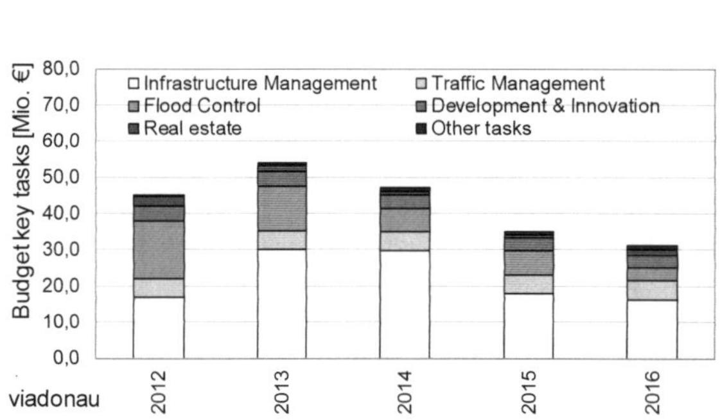

viadonau KEY FACTS:

- maintains 350 km of waterways
- ensures the maintenance of 500 km of towpaths
- manages 200 km of flood protection dams
- maintains 800 km of river banks
- manages about 600 properties
- handles about 100,000 vessels passing through Danube locks per year
- operates the DoRIS (Donau River Information Services)
- 260 employees in headquarter, locks and service centers

Figure 2: Investments in IWW in Austria and key facts on the waterway company viadonau

2. Overview Waterway Asset Management System WAMS

2.1 WAMS – Modules, availability concept and Levels of Service (LOS)

Effective waterway management with sufficient fairway width and depth throughout the year is crucial for an economically successful development of waterway goods transport. Therefore, viadonau decided in 2012 to move from existing empiric reactive approaches towards an analytic proactive life cycle costing approach with the development of a waterway asset management system WAMS (Hoffmann, M. et. al. 2014a,b, Haselbauer, K. 2016). At the core of the developed approach is the total availability concept based on all possible available combinations of fairway width and depth in days per year with the levels of service accounting for typical encounter situations at a minimum fairway depth of 2.5 m (Fig. 3).

The principal methodological approach and software WAMS 1.0 with Module 1: Database, core system and availability and Module 2: Dredging Management have been developed in cooperation with Vienna University of Technologies - Institute of Transportation from 2012 to 2015 with the software being operational since July 2015. In the second stage viadonau teamed up with the principal developers (Hoffmann, M. & Haberl, A.) in order to extend the system with Module 3: Sediment Management, Module 4: Structure Management and Module 5: Traffic Management (Fig. 4) from 2016 to 2018.

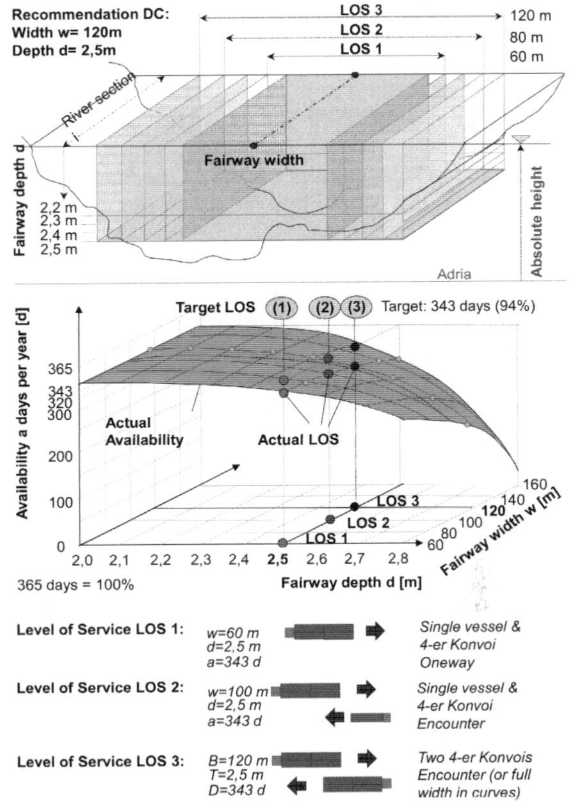

Figure 3: Fairway recommendations, total availability concept and Levels of Service

The focus of this paper is presenting an overview on principal considerations and methods together with developed functionalities and results from Module 3: Sediment Management in the WAMS software. The goal of Module 3 is to provide the waterway operator viadonau with all necessary information regarding sedimentation and erosion processes on the free-flowing sections of the Danube in Austria. In the following sections, the paper describes the basic approach in sediment management.

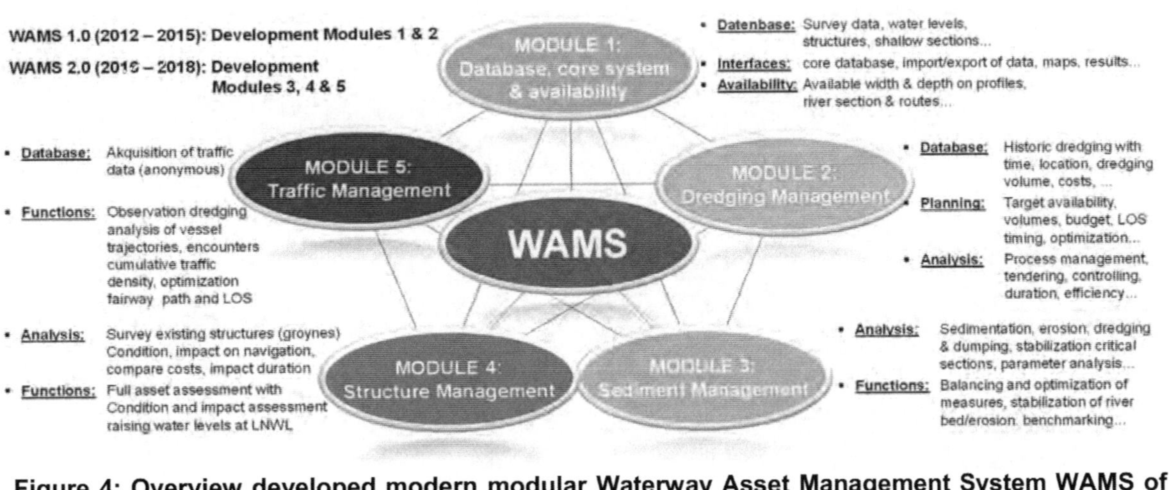

Figure 4: Overview developed modern modular Waterway Asset Management System WAMS of viadonau being operational since 2015

2.2 WAMS – Module 1: Database, core system and availability

Starting from a holistic asset management and total availability perspective, trying to achieve recommended fairway parameters under changing conditions with limited budget cannot yield efficient results. Instead, calculating all possible combinations of available fairway width and depth yields a concave falling availability surface with any target condition being just a point above or below this surface. Improving insufficient availability (e.g. fairway alignment, maintenance dredging, river engineering) starting with the most critical section will lead to increasing costs for additional availability. Drawing from previous research on transport costs it can be shown that increasing fairway depth at low water periods even at the expense of narrowing fairway width is cost efficient, as transport costs are mainly related to actual draughts loaded on a transport route. Optimizing for a given budget therefore yields the combination of width and depth with minimal transport costs (Hoffmann, M. et. al. 2014a,b).

For a calculation of fairway availability, it was necessary to manually assess all shallow sections based on data from riverbed surveys (single & multi beam) in combination with water levels calculated from gauging stations and actual fairway alignment. Apart from the massive amount of data from riverbed surveys alone any manual assessment would take too much time to be efficient, also preventing a transition from reactive maintenance to quantitative analysis and pro-active waterway asset management. Therefore, the development of the WAMS system started with Module 1, acting as a central SQL database and graphical interface for analysing riverbed surveys, gauging data, fairway alignment and marking. Based on this, Module 1 is capable of analysing surveys, water depths, sedimentation and erosion processes, providing the user with automated maps, charts and recommendations. Furthermore, Module 1 combines this information to calculate fairway availability for any given set of parameters as well as availability surfaces for every cross-sectional profile, shallow section or the entire Danube in Austria (350 km) for any given time frame. Additional capabilities are an early-warning system and priorities regarding critical developments (Fig. 5).

Apart from the principal capability of storing and analysing data the main requirement for developing Module 1 was streamlining the entire process of data acquisition, verification, analysis and output based on a user-friendly design, reducing processing time from two to three weeks to just a few days. In a next step, the released capacities were used to import available data from previous years providing information on systematic developments and correlations. Furthermore, riverbed survey intervals were shortened on most critical sections, allowing to provide inland navigation with most recent information on fairway availability as well as gaining further insight into the efficiency of conducted maintenance measures. Setting up a backup-system as well as consistent user permissions from limited view (guest), data input (user) and administrator ensures data safety and integrity in each step of the process. In summary, Module 1 acts as central database, platform and viewer for all other modules of the software.

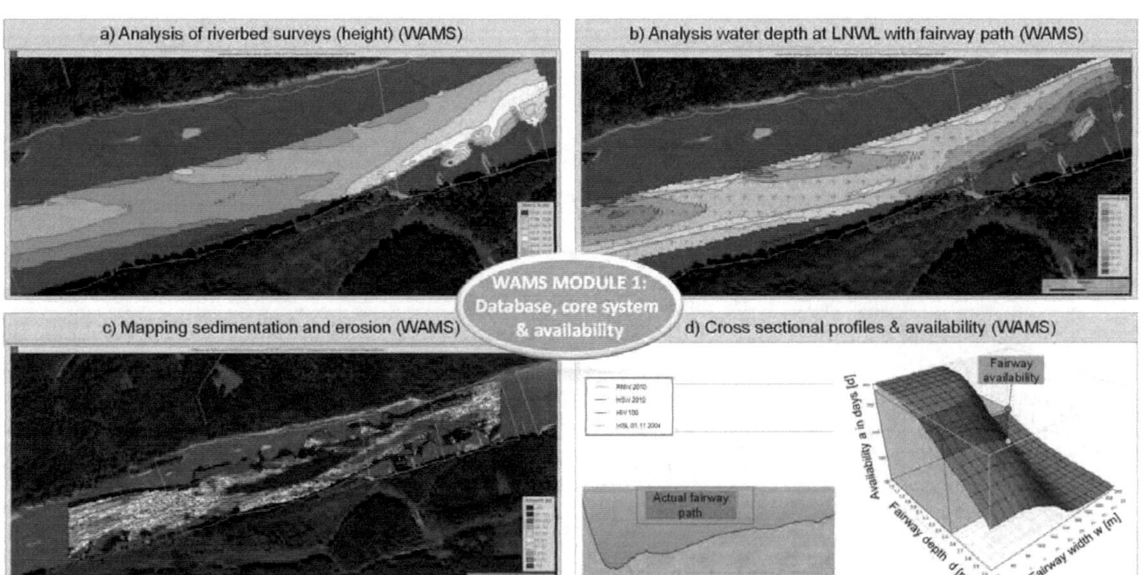

Figure 5: WAMS Module 1 with automatic analysis of a) Riverbed survey, b) Water depth, c) Sedimentation & erosion, d) Cross sectional profiles & availability calculation

2.3 WAMS – Module 2: Dredging Management

Prior to the implementation of WAMS Module 2, fairway maintenance dredging measures had been based on a manual assessment of shallow sections, a rough estimation based on the profile method of possible dredging volumes and a cost estimation derived from average unit costs of previous years. In a next step, selected and approved dredging measures were tendered with a riverbed survey before and after dredging for assessment and billing purposes (Fig. 6). This kind of "manual" dredging planning, which is based on experience, was time-consuming and not capable of providing systematic analysis on alternatives. Furthermore, dredging volume and cost estimations took time, were of limited accuracy and did not provide information on impact duration as a basis for life-cycle analysis and management.

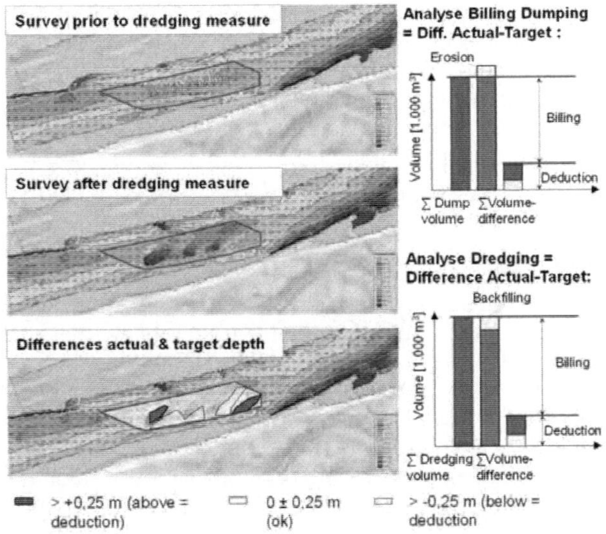

Figure 6: Calculation of dredging and dumping volumes controlling for deviations due to backfilling and differences to target depth

WAMS Module 2 reproduces the entire dredging process, starting from an automated analysis of critical sections, planning of dredging measures per drag and drop, automated cost estimations based on economy-of-scale cost functions and an overview on the status of all measures (Fig. 7). Prior to the development of the WAMS software, estimations of dredging volumes had been based on single-beam surveys and the profile method or a more accurate "manual" assessment of multi-beam data in ArcGIS. In Module 2 both methods are implemented for automatic analysis, revealing substantial deviations in comparison. However, as the analysis with multi-beam riverbed surveys is more accurate, it is now standard for deviation analysis in billing of conducted dredging measures. Further functionalities of this module include an analysis of dredging impact duration based on sedimentation and erosion rates, enabling an assessment of the efficiency of a dredging measure in comparison to other solutions (e.g. fairway alignment, construction of groynes). In summary, the development of Module 2 allows much faster and efficient planning, implementation and controlling of dredging to achieve higher fairway availability. The comprehensive analysis and documentation system is constantly being updated based on previous results, thus becoming more accurate with every year. Since 2015, Module 2 has been successfully used for planning of all dredging measures, providing a constant feedback for improvements.

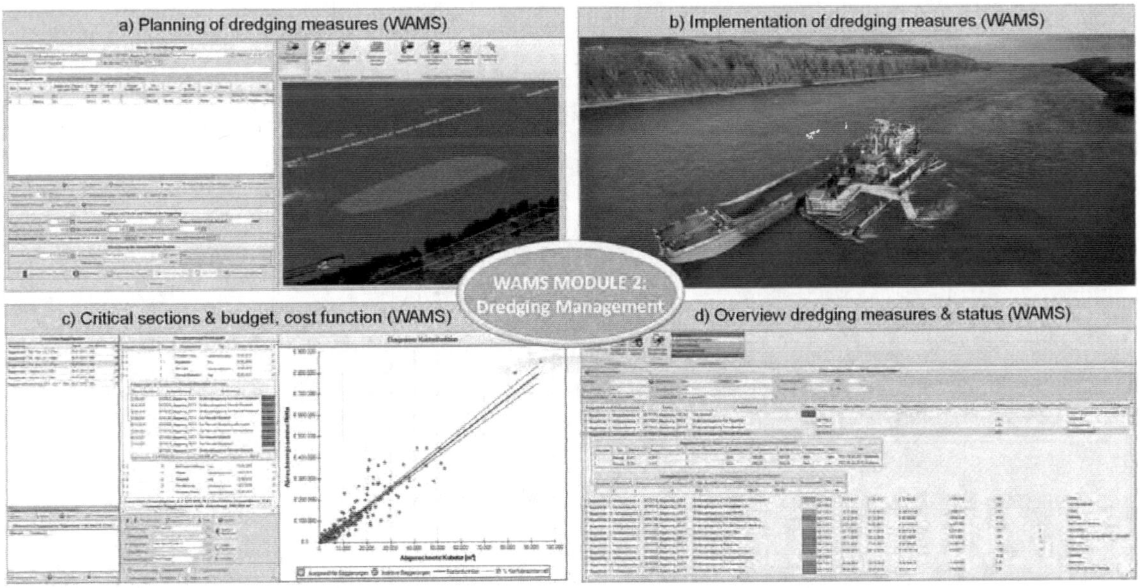

Figure 7: WAMS Module 2 with a) Planning of dredging, b) Implementation & controlling, c) Critical sections, costs and budgeting, d) Overview status and implementation of dredging

2.4 WAMS – Module 3: Sediment Management

The river Danube in Austria features a number of consecutive hydraulic power plants and two major free-flowing sections with a number of shallow sections from Melk to Krems and from Vienna to the Austrian-Slovakian border. Parts of these sections are characterized by river-engineering structures (groyne fields) with substantial impacts on river morphology. Due to substantial stretches of the river banks being fixed and a dynamic riverbed, there have been ongoing erosion tendencies between Vienna and Hainburg of roughly 1.0 m from 1950 to 2010 (Klasz, G. 2013, 2015). Furthermore, there are a number of scours showing a low thickness of stable riverbed layer which require regular monitoring with regard to critical erosion tendencies.

Thus, the goal for Module 3 was to develop and implement a concept for a systematic assessment of critical scours, assessing sedimentation and erosion rates in these areas. Furthermore, to provide analytic capabilities, an automatized overview has been implemented on all conducted dredging and dumping activities countering erosion together with a volume balance sheet covering historic and actual developments. The developed approach is based on dredging shallow sections and dumping dredged material in critical scours upstream, thus improving fairway availability while countering riverbed erosion at the same time. Another necessary analytical capability of Module 3 is a detailed analysis of shallow sections with an assessment of backfilling behaviour after dredging measures depending on water levels and discharge. The same goes for the analysis of dumping sites and critical scours, with this kind of analysis providing the basis to distinguish between “natural erosion” and impact of dredging with dumping upstream.

Based on the concept for a systematic sediment management (Klasz, G. 2015, Hoffmann M. et. al. 2016), Fig. 8 provides an overview on already implemented software features. Module 3 already facilitates an analysis of sedimentation and erosion rates on short sections for a fixed time frame as well as the development over time including dredging and dumping measures. Further capabilities are an overview on all conducted dredging measures and related dumping sites for a given time frame as well as a total balance on all erosion and sedimentation volumes for any given time frame and river stretch (sediment balance).

From a waterway asset management perspective, Module 3 provides viadonau with the capability of going beyond optimal fairway parameters for inland navigation, balancing necessary maintenance and the needs of environmental protection and stabilization of the riverbed. Instead of lengthy analyses and studies this module allows viadonau to constantly assess and adapt their approaches by optimizing both the selection and the timing of appropriate measures. In summary, Module 3 will help to efficiently balance the interests of environmental protection and inland navigation. Furthermore, the generated results will provide viadonau with the basis for future insights and improvements.

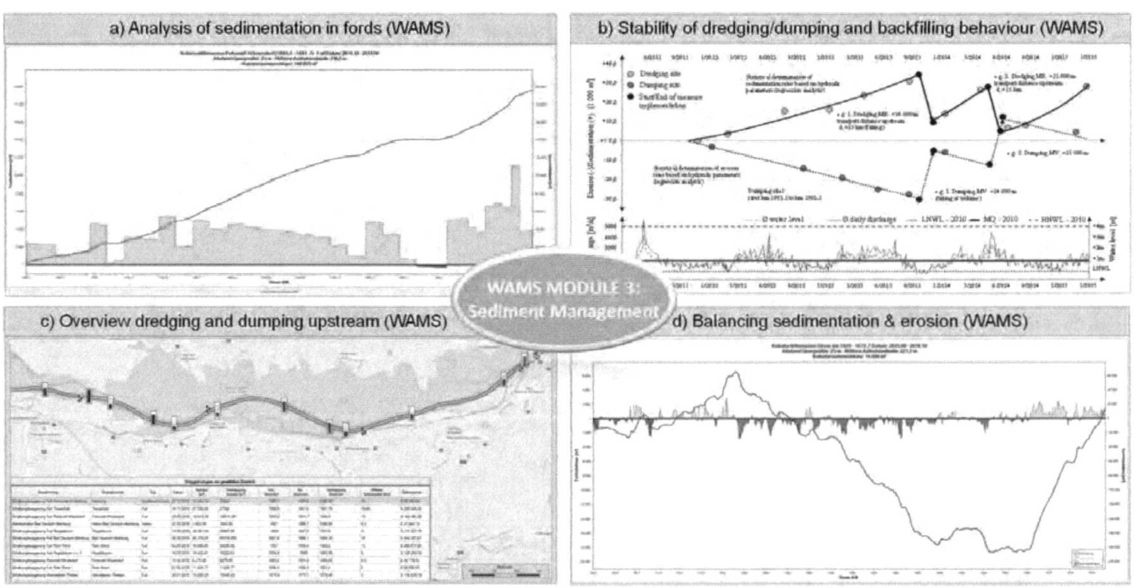

Figure 8: WAMS Module 3 with a) Sedimentation in fords, b) Stability of dredging/backfilling, c) Overview conducted dredging and dumping, d) Balancing sedimentation and erosion

2.5 WAMS – Module 4: Structures Management

On the free-flowing section of the Danube east of Vienna there are a large number of hydraulic-engineering structures (groynes, training walls) as well as some embankment and bottom protection measures which were mainly constructed at in the 20th century. In Austria groynes are constructed from loose boulders reducing cross sections and thus increasing water levels and as well as fairway depths at low navigational water levels (LNWL). In contrast, airway dredging the fairway path lowers the river bed increasing also increases fairway depths, but as well also slightly lowering lowers water levels (Fig. 9). Thus, dredging for greater fairway width becomes increasingly inefficient when compared to long- lasting groynes, especially if dredging exhibits a short impact duration due to high sedimentation rates (Krouzecky, N. 2015, Hoffmann, M. et. al. 2016).

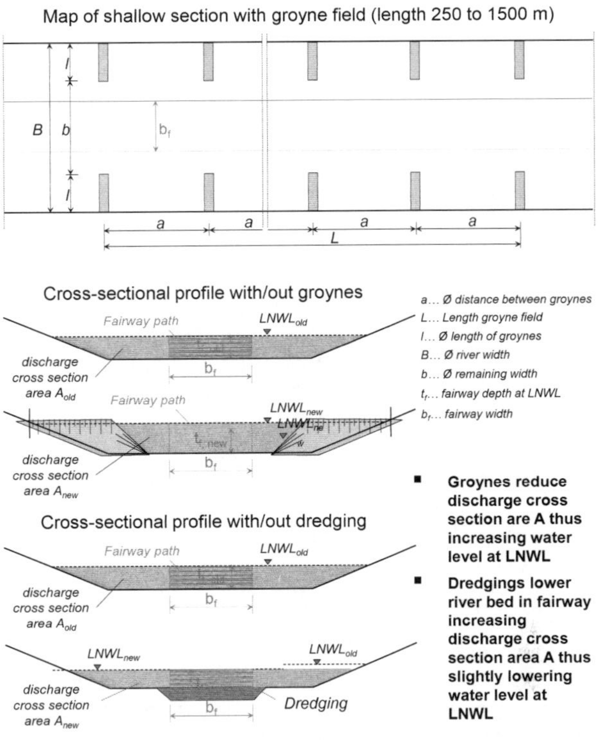

Figure 9: Impact of river engineering measures and dredging in the fairway path on available depth at low navigational water levels (LNWL)

The goal of in developing Module 4 therefore was therefore to develop provide a consistent approach regarding an implementation of waterway structures in the waterway asset management as a basis for decision making. The developed functionalities of the WAMS in this module include a survey of existing and/or defective waterway structures, volumetric analysis for assessing possible reconstruction costs as well as a functional assessment regarding fairway availability. Based on an overview and analysis of existing waterway structures and their functionality, it will be possible to decide whether dredging measures or lasting river engineering are cost efficient. Furthermore, this will allow to minimize necessary interventions, balancing the environmental needs with those of inland navigation. Furthermore, acting as a basis for planning and cost estimations in the reconstruction and adaption of existing or necessary additional waterway structures. In summary, Module 4 will enable viadonau to include hydraulic-engineering structures in the holistic asset management process of WAMS.

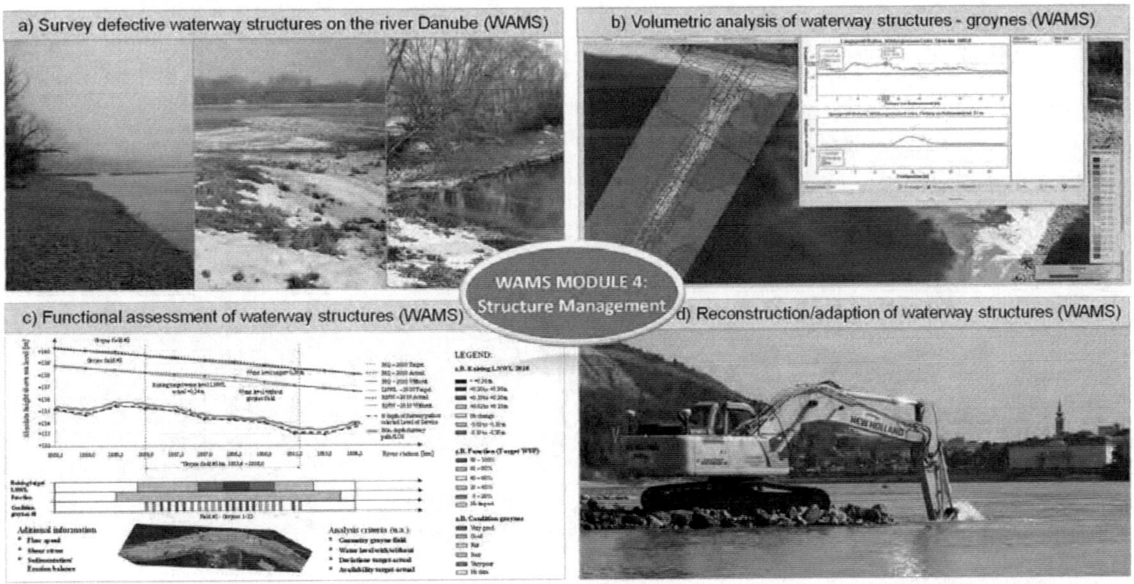

Figure 10: WAMS Module 4 with a) Survey defective structures, b) Volumetric analysis, c) Functional assessment of structures, d) Reconstruction and adaption of waterway structures

2.6 WAMS – Module 5: Traffic Management

As, previously mentioned, the fairway availability on a transport route is always limited at a certain time by the most critical shallow section, a non-functioning lock or other physical impediments. With two locks on every hydrological power station as well as an almost permanent availability of at least one lock with sufficient capacity for usual traffic volumes, the main issue is providing a sufficient fairway depth of at least a deep fairway channel. Thus, shallow sections form a serial system with changing availability over time, with the total availability being defined by the critical path for any given time frame. As necessary maintenance and especially river engineering measures take time, it is of crucial importance to determine critical developments prior to their occurrence based on comprehensive prediction functionalities. Other Another possibility is a fairway alignment which is much faster in implementation but is limited by river width, curvature and actual traffic flows. For an optimization of the efficiency of both waterway transport and efficiency of communication results it is also important to know actual of availability the current utilization of fairway availability based on a comprehensive analysis of traffic flows. The goal for the development of Module 5 was therefore to extend the WAMS with additional capabilities regarding actualization of fairway paths, markings and buoys as well as an integration of all important navigational structures on river shores. Further functionalities are tracking dredging vessels between dredging and dumping sites, integration and analysis of anonymized vessel trajectories for traffic and encounter analysis. However, the main goal was a quantitative assessment of traffic flows and distribution on cross- sectional profiles and maps with the possibility to align the fairway path, Levels of Service (LOS) and signalization & marking. In addition, conveying reliable availability information to skippers via signalization marking and different communication channels is equally important for enabling a high utilization rates (Haselbauer, K. 2015, Hoffmann, M. et. a. 2016).

The developed WAMS Module 5 is capable of tracking dredging vessels as they are obliged to provide this information for controlling purposes based on dredging contracts. In contrast, an automated import function for track information from passenger and goods transport uses an automated import function to anonymizes transponder data, leaving computing only vessel type, position and draught loaded. The developed algorithm for an analysis of traffic flows is based on the intersection of vessel trajectories with a defined grid being used as counting cells. Possible vessel encounters can be calculated from vessel position over time being projected to actual river kilometre. In turn, this allows to map specific encounter situations and distances being of specific importance regarding with regard to fairway alignment in narrow sections. Finally, the cumulative traffic density being represented as heatmaps and distribution in cross- sectional profiles (with average and confidence interval) enable an efficient alignment of both fairway and levels of service (Fig. 11). In summary, Module 5 provides viadonau with the capability to optimize waterway availability and fairway path according to actual traffic flows. Based on the results viadonau is also capable of providing actual information to the shipping industry regarding unused availability and loading potentials in order to improve navigation's competitivety on the market.

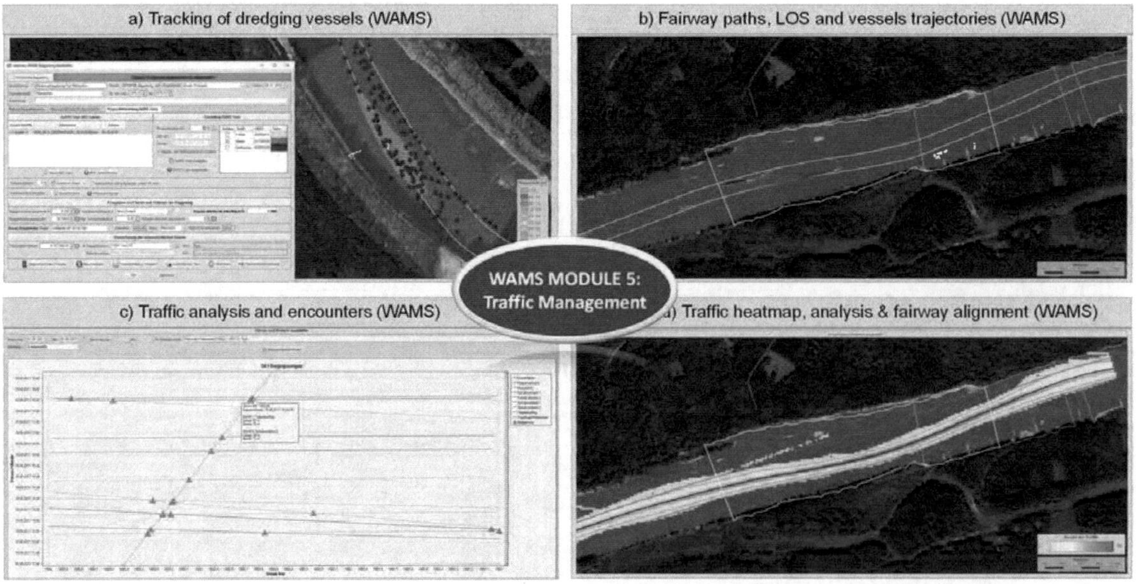

Figure 11: WAMS Module 5 with a) Tracking of dredging measures, b) Fairway paths and levels of service, c) Traffic analysis and encounters, d) Traffic heatmap, analysis & fairway alignment

3. Methods and approaches in Sediment Management (Module 3)

3.1 Assessment of critical fords

The Dredging Management (Module 2) includes the capability to calculate dredging volumes and costs in shallow sections and fords based on single and multi-beam surveys for any defined area and target depth. Surveys of fairway conditions before and after dredging measures allows to check whether targeted depths have been achieved and actual dredging volumes are consistent with billed volumes. However, full life cycle capabilities regarding dredging measures require an assessment of impact duration with backfilling rates on dredging sites and erosion on dumping sites. With Module 3: Sediment Management an automatic analysis of typical sedimentation and erosion rates together with a full-scale documentation of measure implementation becomes feasible (Hoffmann, M. et. al. 2014, 2016; Haselbauer, K. 2016).

Figure 12 provides an overview on the assessment of dredging and dumping impact duration based on a calculation of volume deviations between consecutive surveys either with the cross-section profile-method (single-beam) or a full calculation with TIN-networks (multi-beam). On typical fords there will be an ongoing sedimentation with increasing volume accumulation until implementation of dredging measures. At scour areas there will be an ongoing erosion process that can be mitigated by dumping dredged material from fords. By systematic comparison of discharge levels, flow speed and water levels with changes in the riverbed an empiric calibration of sedimentation and erosion rates for different conditions becomes feasible.

As sedimentation and erosion are cumulative processes these empirical calibrated rates will be a good basis for the prediction of these processes as well. As multi-beam surveys on critical fords or scour areas are conducted at least every 1-2 months the results of future prediction capabilities in the WAMS can be adjusted accordingly. Including these functionalities in the WAMS enables the waterway operator to assess the efficiency of dredging measures for individual fords compared to other e.g. more permanent solutions (e.g. adaptation/construction of groynes). Based on these results it is possible to optimize extent and timing of dredging measures, enabling future condition prediction and stable budget forecasting results. Thus, the implementation of these functionalities will ensure an efficient use of funds and high availability of the waterway.

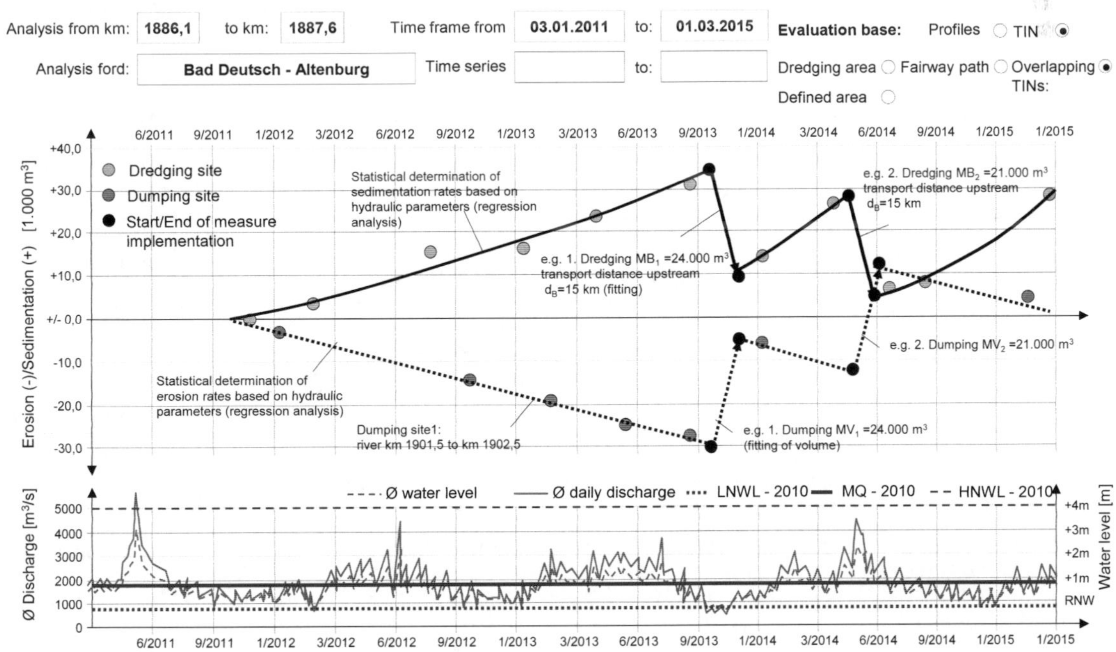

Figure 12: Dredging and dumping impact duration with analysis of typical sedimentation and erosion rates from analysis of periodic multi-beam surveys and dredging/dumping measures

3.2 Assessment of critical scour areas

The river Danube in Austria has a number of consecutive hydraulic power plants and two major free flowing sections in Wachau and east of Vienna down to Hainburg at the border to Slovakia. At these sections the Danube is dominated by natural stretches as well river engineering structures (groyne fields) with substantial impact on river morphology. Thus, long stretches of river banks together with the principal river path are fixed whereas the riverbed remains flexible and malleable. As a result of this river regulation and lack of bed-load there has been an ongoing erosion mainly on the river section from Vienna to Hainburg of roughly 1.0 m from 1950 to 2010. In addition, there are certain scour areas with a low thickness of stable river bed layer and critical erosion rates (Klasz, G. 2013, 2015).

The concept for an assessment of critical scour areas based on thickness of stable layer, sedimentation and erosion rates with systematic dredging in sedimentation areas and dumping in scour areas up-stream with volume balance in longitudinal profiles is described in Figure 13. The basic idea is a systematic backfilling of scour areas upstream with material from dredging measures on critical fords down-stream. The layer thickness of the stable layer is determined based on core-sampling on both sides of the river and at selected positions in the riverbed in critical sections. A systematic analysis of changes in altitude of the riverbed on river sections provides insights into sedimentation and erosion rates. The basis for this analysis are bi-annual multi-beam surveys of the entire free-flowing sections. With this it is possible to detect scour areas with the need to counter critical erosion processes.

With Module 3: Sediment Management it will be possible to determine possible scour areas and critical erosion tendencies automatically providing viadonau with automated warnings if the thickness of the stable layer falls beyond a given threshold. However, from current analysis it can be concluded that the number and extent of critical scour areas in the free-flowing sections of the Danube in Austria is limited despite general erosion tendencies. In addition, current dredging volumes at shallow sections exceed the necessary amount of material for stabilizing these sections. With regular monitoring and systematic dredging and dumping in critical scour areas it is therefore possible to avoid critical developments. If the dredging material is not stable enough it is also possible to counter erosion by dumping additional material with an optimized grain-size distribution and monitor further developments with the WAMS.

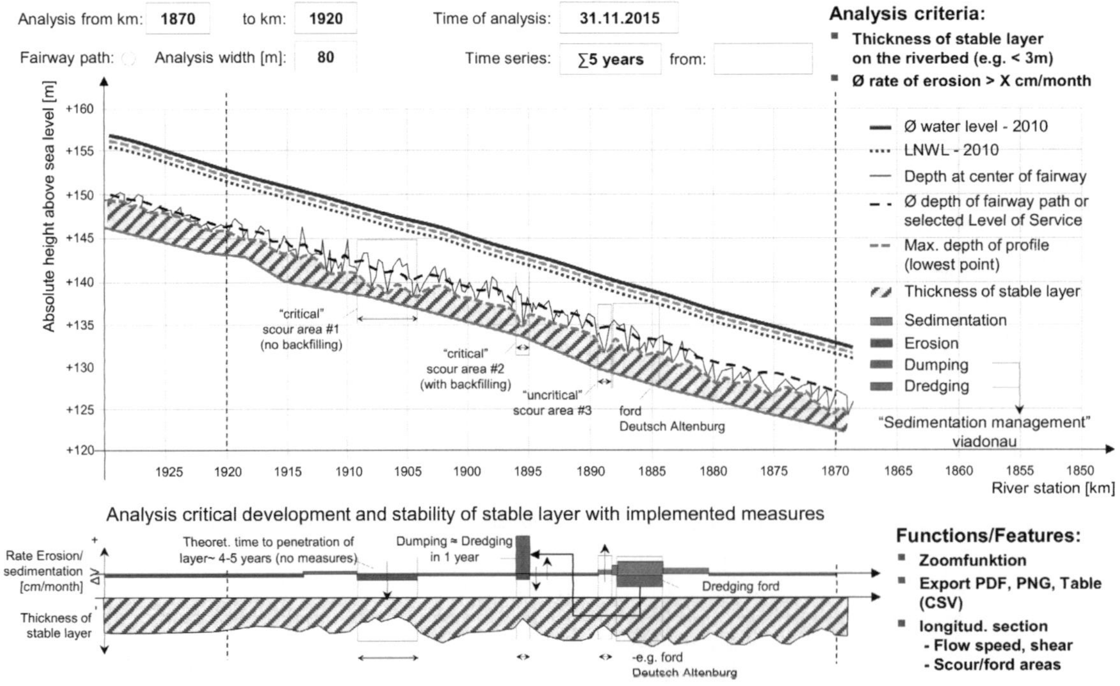

Figure 13: Assessment of critical scour areas based on thickness of stable layer, sedimentation and erosion rates with systematic dredging in sedimentation areas and dumping in scour areas

3.3 Balancing erosion and sedimentation

As previously mentioned the river Danube exhibits certain erosion tendencies in the free-flowing sections due to regulations and a lack of bedload below hydro-electric power plants despite sedimentation on a few shallow sections mainly east of Vienna close to the border of Slovakia. On this short river stretch of roughly 50 river-kilometer the erosion lead to a loss of 2,2 Million cubic meter of bedload from 1996 to 2015 translating in several decimeters of riverbed erosion. Between 1950 to 2015 the riverbed erosion exhibited a progressive growth exceeding 1,2 m in average according to extensive analysis. Apart from the impact on technical structures and the ground water level this development is also critical for the environment with its protected meadows and flood plains (KLASZ, G. 2013, 2015).

Figure 14 provides a concept for an overview and volume balancing in sediment management based on natural erosion tendencies, dumping of addition-al material below a main hydroelectric power plant (Freudenau - VHP) and all dredging measures in sedimentation areas with dumping upstream. Based on actual surveys it is possible to determine a mass equivalence from dredging and dumping upstream depending on transport distance that can be calibrated based on actual data. With this it is possible to calculate cumulative "natural" erosion volumes for an analyzed river section without measures as a baseline being countered to some extent by additional material from VHP and sedimentation management. Minimizing total erosion volumes to acceptable levels either based on additional dumping material up-stream or sedimentation management (e.g. increase in transport distance or dredging volumes) will be subject to an iterative process.

However, implementing these functionalities in WAMS – Module 3: Sediment Management allows viadonau to keep track of all implemented measures and their impact providing the necessary information for decision making on an actual basis. Thus, these functionalities act as an instrument providing information on critical developments on scour and sedimentation areas allowing for timely interventions together with accurate information on necessary extent and costs. In addition, this will provide evidence regarding treatment impact being a basis for optimization. Regarding high level decision making the automatic analysis of achieved results with the given budget and selected measures will provide the necessary evidence for further improvements.

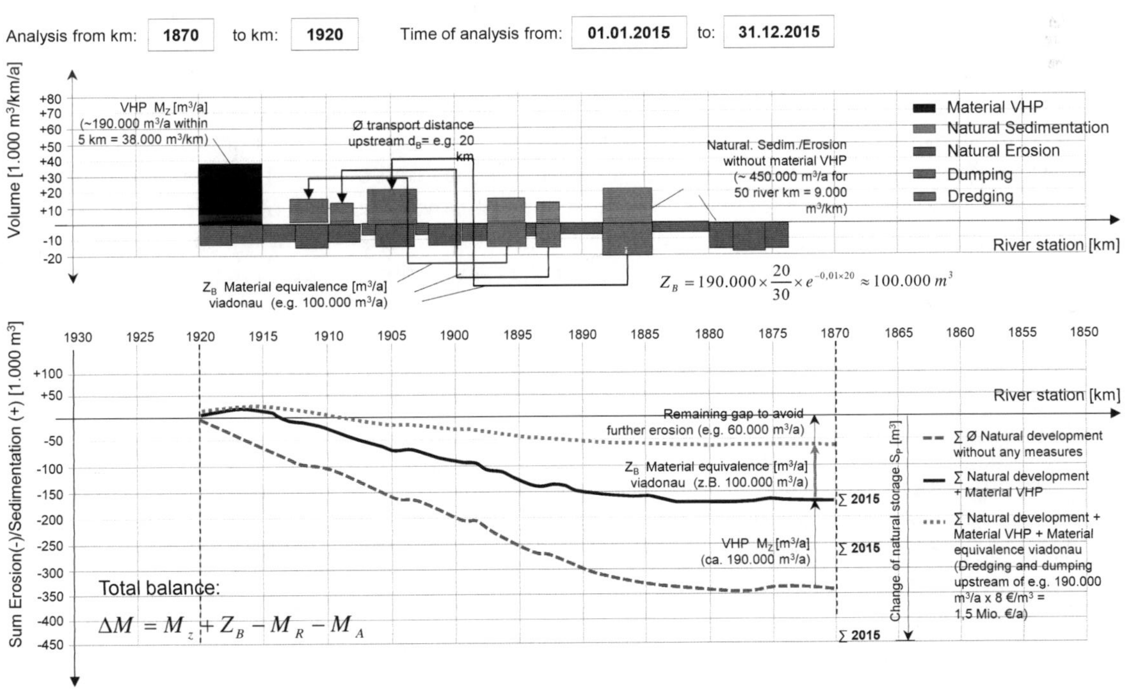

Figure 14: Volume balancing in sediment management with natural erosion, dumping below a hydro-power plant (VHP), dredging in fords and dumping upstream with mass equivalence

4. Conclusions and outlook

For managing a dynamic river as transport infrastructure in a cost-efficient and environmentally friendly way viadonau has teamed up with Vienna University of Technology and Hoffmann Consulting in order to develop a holistic Waterway Asset Management System (WAMS). In a first phase from 2012 to 2015 the principal methodological approaches for Modules 1: Database, core system and availability and Module 2: Dredging Management have been developed (WAMS 1.0). In the second phase from 2016 to 2018 Module 3: Sediment Management, Module 4: Structure Management and Module 5: Traffic Management are being implemented (WAMS 2.0). With the WAMS in operation since July 2015 the development is work in progress providing constant feedback between theoretical considerations and practical results as new functionalities become available. In this regard, the WAMS is becoming an integrator of all kinds of data providing viadonau with the means to move from empiric reactive maintenance approaches towards quantitative asset management strategies with fast semi-automated processing capabilities and pro-active maintenance in a user-friendly environment.

With the introduction the paper provides an insight into investments in inland waterways both in OECD – Countries and Austria. In the second part of the paper an overview of the WAMS – Modules and their functionalities together with the availability concept and implemented levels of service (LOS) are given. The third part of the paper presents further insights into the approaches on concepts behind Module 3: Sediment Management starting with the assessment of critical fords. Based on a comprehensive analysis of sedimentation in fords and erosion in scour areas the stability and impact duration of dredging and dumping measures are analysed. By systematic comparison of discharge levels, flow speed and water levels with morphological changes this functionality will yield empirical sedimentation and erosion rates allowing for a better planning and prediction of treatment impact. The assessment of scour areas together with the thickness of stable riverbed layer provides an insight into critical developments. Based on a thorough analysis it can be shown that critical developments can be mitigated by systematic monitoring and dumping dredging material in critical scour areas.

Regarding sediment management apart from the development of fairway conditions one of the most critical developments on the free-flowing sections has been the constant erosion process in the last decades. From 1950 to 2015 alone the riverbed erosion in the free-flowing section south of Vienna led to a riverbed erosion exceeding 1,2 m in average despite some dumping efforts below a hydro-electric powerplant at the beginning of this river stretch. This negative bed-load balance is especially problematic for protected natural habitats and their flora and fauna with meadows and flood plains falling dry more frequently. By implementing Module 3: Sediment Management the impact of systematic measures countering these erosion processes e.g. by dredging downstream in fords and dumping upstream can be monitored and optimized. The automated analysis of achieved results with different measures not only provides evidence that these erosion processes can be limited but also allows for more accurate measure planning and timing in the future. In summary, the WAMS – Software is a tool providing decision makers in viadonau with the means for a holistic view on the entire waterway Danube allowing balanced and efficient investments.

Acknowledgements

Developing the waterway Danube towards a smart transport infrastructure requires the effort of all stakeholders. The presented results would not have been possible without funding from EU, BMVIT and the collaborative effort of the WAMS project team (TU Vienna – Institute of Transportation, Hoffmann Consulting & viadonau).

References

Haselbauer, K., 2016. Transport Infrastructure Asset Management. PhD-Thesis Vienna University of Technology, Vienna

Hoffmann, M., Haselbauer, K., Haberl A., Blab R., Simoner M., Dieplinger K., Hartl T., 2014. A new asset management approach for inland waterways. Peer reviewed proceedings Transport Research Arena TRA2014, Paris

Hoffmann, M., Hartl, T., 2016. Actual developments, treatment options and future waterway asset management implementation for the waterway Danube. Peer reviewed proceedings IALCCE2016, Delft

Hoffmann M., Haberl A., Konzel C., Hartl T., Simon S., Simoner M. (2018); Traffic analysis, fairway alignment and efficient investments in the waterway Danube; Proceedings of 7th Transport Research Arena TRA 2018, April 16-19, 2018, Vienna, Austria

Klasz, G. (2013); Die flussmorphologische Situation der Donau in Hinblick auf Sohlerosion; in German; Issue 29/2013; Nationalpark Donauauen; Vienna.

Klasz, G. (2015); Zu den Möglichkeiten einer Geschiebebewirtschaftung und den zugehörigen Optimierungspotentialen für die Donau östlich von Wien; in German; Issue 37/2015; Nationalpark Donauauen; Vienna.

Krouzecky, N. 2015. Comparative analysis of water level elevation induced by groynes; scientific study, Institute of Hydraulic Engineering and Water Resource Management; Vienna University of Technology; Vienna.

viadonau, 2013. Manual on Danube Navigation. ISBN 978-3-9502226-2-3, Vienna

viadonau, 2016. Annual Report on Danube Navigation in Austria, Vienna

viadonau, 2017. Annual Report on Danube Navigation in Austria, Vienna

Pneumatic Flow Tube Mixing for Processing and Beneficial Use of Fine-Grained Dredged Material

March 2018

Ali Maher, PhD[1]
Robert Miskewitz, PhD[2]
Masaki Kitazume, PhD[3]
Scott Douglas[4]
Abbas Sarmad, PE[5]

This paper discusses the feasibility of using a novel sediment solidification/stabilization technique, the Pneumatic Flow Tube Mixing (PFTM), which has been successfully used in Japan for the last decade in large scale reclamation projects utilizing stabilized soft sediments. The study was conducted by the Center for Advanced Infrastructure and Transportation (CAIT) at Rutgers University and its research partners to evaluate PFTM for stabilization and solidification of soft sediments dredged from the New York/ New Jersey Harbor. The study included a comprehensive laboratory investigation aimed at determining the ideal mix for PTFM stabilization, and a pilot scale demonstration of the PTFM technique in the field. Material stabilized during the demonstration was analyzed via unconfined compression tests, needle tests, laboratory vane shear tests, and flow tests in the laboratory and the cone penetration test in the field. Curing times for the samples were 3, 7, 14 and 28 days. The three cement content mixes were 4%, 8%, and 12% by weight (wet weight). Furthermore, samples of raw and stabilized dredged material were evaluated for total constituent concentrations (SVOCs, Metals, Pesticides, and PCBs) via the SPLP procedure. The results of the field testing program demonstrated that PFTM is an effective and efficient technique for solidification/stabilization of soft contaminated sediments from NY/NJ harbor with unconfined compressive strength values from laboratory and field samples averaging about 75 kPa (8% cement content), with a uniform field mixture quality. Furthermore, the chemical analysis of the stabilized sediments indicated no detectable mass of SVOCs, PCBs, or Pesticides.

Keywords: dredged soil, pneumatic flow mixing method, unconfined compressive strength, applicability, field test

1. Prof. of Civil Engineering, Dept. of Civil and Environmental Engineering, Rutgers University, 100 Brett Road, Piscataway, NJ08854-8014, phone (848) 445-2951, email: mmaher@rci.rutgers.edu.
2. Associate Research Professor, Dept. of Civil and Environmental Engineering, Rutgers University, 100 Brett Road, Piscataway, NJ08854-8014, phone (848) 445-0579, email. Email: rmiskewitz@envsci.rutgers.edu
3. Professor, Tokyo Institute of Technology, Tokyo, Japan. E-mail: kitazume@cv.titech.ac.jp
4. Dredging Program Manager, Office of Maritime Resources, New Jersey Department of Transportation, PO Box 837, Trenton, New Jersey, 08625, phone (609) 530- 4770, email: scott.douglas@dot.state.nj.us.
5. Senior Vice President, AECOM, 20 Exchange Place, New York, NY 10017, phone: (917)-945-9026, email: abbas.sarmad@aecom.com. Paper Presenter at the conference

Introduction

The Rutgers University Center for Advanced Infrastructure and Technology recently completed a study to evaluate sediment amendment using the Pneumatic Flow Tube Mixing Apparatus (PFTM) in NY/NJ Harbor. PFTM is a rapid construction technique for stabilizing, transporting, and placing reclaimed dredged materials, soft soils and mud, which allows for their beneficial use as structural and non-structural fills. It can be used on land or off shore (within rivers and canals) and is especially well suited for the reclamation and processing of contaminated dredged materials from large scale navigational and remedial dredging projects. Prior to the development of this method, stabilization of dredged sediment and soils required special mixing plant vessels or cumbersome on-land processing plants. With the PFTM System, onsite pneumatic conveying capabilities, requiring ordinary equipment and facilities, can be used to simplify the mixing process within a small footprint and process enclosed in a tube to minimize odors. It has also been shown to eliminate the expensive and environmentally problematic disposal of dredged materials while simultaneously providing an economical and effective beneficial use application for structural fill purposes. Developed in the early 2000's in Japan, the method has been used extensively in projects of all sizes and scopes including offshore reclamation for the construction of the Tokyo Haneda International Airport Runway, Osaka International Airport, Nagoya International Airport, and within numerous harbors in Japan (Kitazume and Satoh 2003, 2005, Mizukami and Matsunaga 2015).

Method Description

The PFTM stabilization process is entirely enclosed from the time when it is loaded until the stabilized material is discharged from the system. The system receives dredge materials similar to other stabilization systems; a scow containing dredge material is brought to the process location. The scow is unloaded via excavator and loaded into a grizzly.

Figure 1. (a) Vibrating with being loaded from a truck (b) Screen and vibrator configuration.

Once the dredged material is screened it is fed into a hopper where it enters the PFTM process flow. It is pumped as slurry from this hopper to the mixing tool. During transport the slurry density is measured, this measurement is fed to the automated cement pumping system to ensure that the proper dosage of cement is added. A process flow diagram is presented as Figure 2.

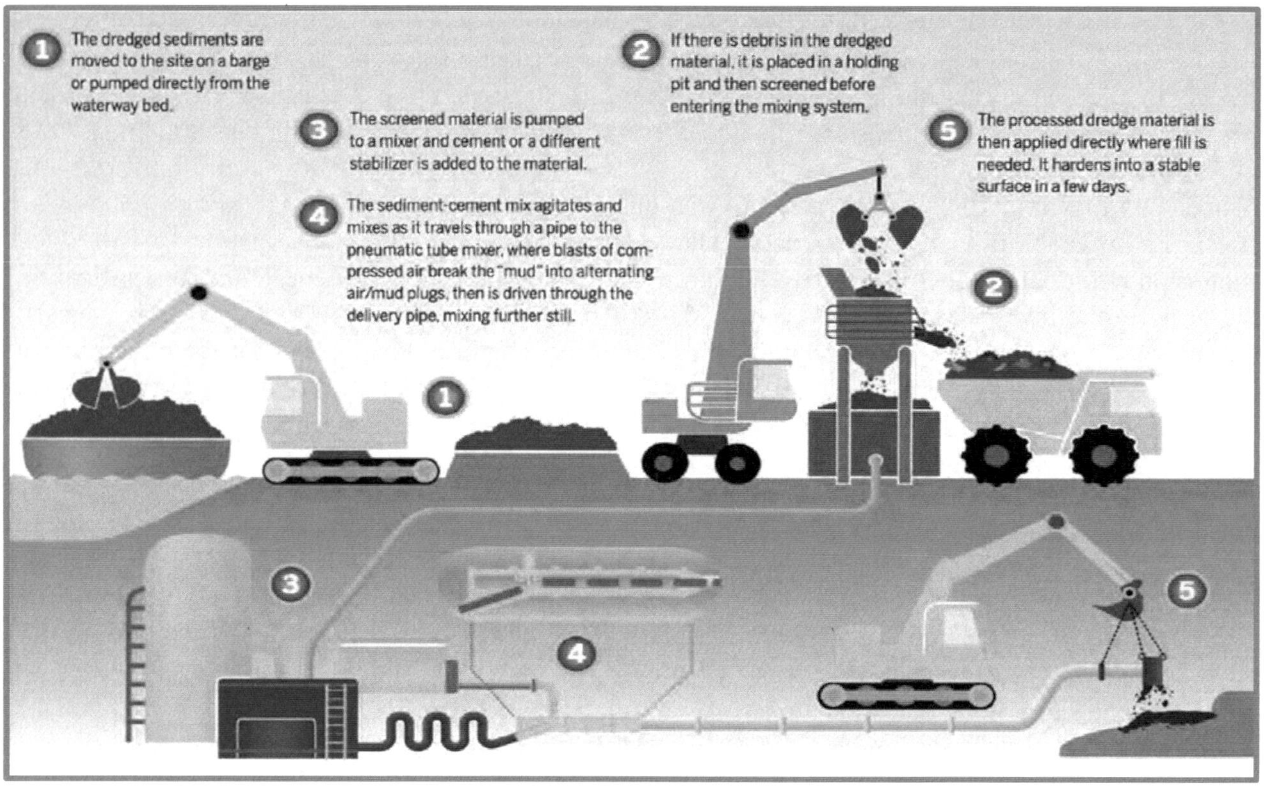

Figure 2. Process Flow Diagram for the PFTM System

The PFTM system mixes sediments with designated binders in a pipe via compressed air, during transportation from a source to the final placement site (Ministry of Transport, The Fifth District Port Construction Bureaus 1999; Kitazume and Satoh 2003; Coastal Development Institute of Technology 2008). The mixture of soft sediment and binder (cement) forms many separated mud-plugs in the pipe, and these are thoroughly mixed during transport via turbulent flow (short distance) generated within the plug. The soft sediment stabilized with binder has rapid increase in strength, and the strength of stabilized sediment can be easily controlled by changing the amount of binder and water content of the soil through a real-time computer monitoring system. The sediment mixture deposited and cured at the site can gain relatively high strength rapidly so that no additional sediment/soil improvement is required. Transport of soft sediment in a pipe without any amount of air requires high pressure to compensate for friction generated along the inner surface of the pipe. When pulses of compressed air are injected into the pipe together with soft sediment and the binder, the mixture is separated into small plugs. The plugs are then pushed to an outlet. The formation of plugs reduces the friction along the pipe's inner surface and can considerably reduce the air pressure required for transport.

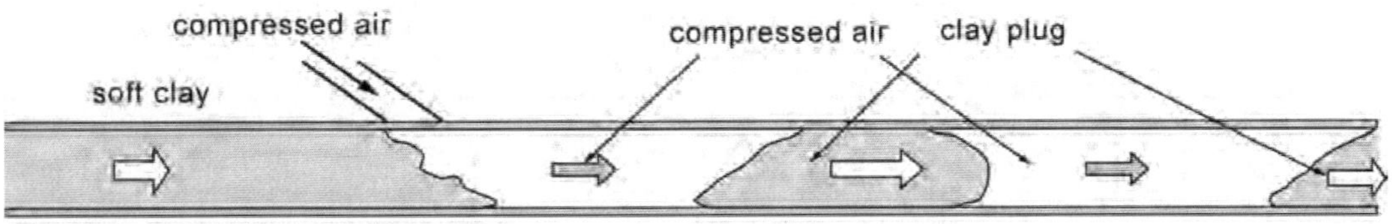

Figure 3. Conceptual Diagram of Mud-Plugs

The air pressure required for transporting sediment depends upon many factors such as sediment properties, injected air volume, pipe diameter and length. In the current practice, an inlet air pressure of 400 to 500 kPa is used (Coastal Development Institute of Technology 2008). The plugs are transported at speeds exceeding 10 m/sec in the pipe resulting in turbulent flow within the plugs due to the friction along the inner pipe surface which mixes the sediment and binder. PFTM has been used for a variety of sediment types with varied effectiveness. The results are summarized in Table 1.

Table 1. Soil properties suitable for pneumatic flow tube mixing method (Coastal Development Institute of Technology, "Technical Manual of Pneumatic Flow Mixing Method", revised version, 187p, 2008.)

Soil type		Water content	Applicability	Evaluation	
				binder content	Pumping ratio
Cohesive soil	Sand content 30-50%	High	good	middle	Somewhat low
		Low	good	middle	Low
	Sand content < 30%	higher than 200% (higher than 2.8 wl)	not good	-	-
		110 to 200% (1.5 to 2.8 wl)	good	rich	fair
		90 to 110% (1.3 to 1.5 wl)	good	middle	fair
		70 to 90% (1.0 to 1.3 wl)	good	middle	Somewhat low
		50 to 70% (0.7 to 1.0 wl)	good	middle	Low
		lower than 50% (lower than 0.7 wl)	not good	-	-

Plugs of stabilized material are transported to the discharge point. This discharge can be located at a staging area or potentially at the point of use, such as for filling behind bulkheads.

Figure 4. (a) PTFM mixing pipe (b) discharge of stabilized material from pipe

NY/NJ Harbor Demonstration Project

The evaluation of the PFTM method for use in the New York / New Jersey Harbor was conducted at a sediment management site in Kearny, New Jersey operated by Clean Earth Inc., of Hatboro, Pennsylvania, from July through September of 2015. The PFTM system was brought from Japan and erected at the site. The first preliminary test occurred on 7/24/2015, and the first week of production was 8/4/2015 through 8/7/2015. The total amount of material processed during the deployment was limited to 3929 m^3. The field production was completed on 9/4/2015. This project successfully demonstrated that the PFTM process could be used for impacted sediments dredged from the Arthur Kill waterway - NY/NJ Harbor and the resulting amended material possessed the required strength and chemical characteristics for upland placement and beneficial reuse.

Chemical analyses completed for these samples include SVOCs, Metals, Pesticides, and PCBs. Each of the three raw materials were analyzed for total constituent concentrations (mass of contaminant/mass of sample) and for leachate concentration via the SPLP procedure. Comparison of the total constituent concentrations to the NJ Residential Direct Contact Soil Remediation Standards indicates that the material is relatively clean with few violations. Violations included Benzo(a)anthracene, which ranged in concentration from 0.171 to 1.79 mg/kg (standard = 0.6 mg/kg), Benzo(a) pyrene which ranged in concentration from Not Detected to 1.24 mg/kg (standard = 0.2 mg/kg), and Benzo(b) fluoranthene which ranged in concentration from 0.182 to 1.85 mg/kg (standard = 0.6 mg/kg). In addition, all of the materials had elevated arsenic concentrations (average value of 40 mg/kg with a range ±5 mg/kg) above the 19 mg/kg standard.

Analysis of the contaminants in the leachate derived from the SPLP procedure indicates that there is no detectable mass of the target SVOCs or any of the other SVOCs compounds. The concentrations of Arsenic in the leachate from the raw sediment for the 8% mix were 14.8 and 15.9 µg/L and 3.97, 3.47 and 3.97 µg/L in the stabilized material. Similarly, the concentrations of Arsenic in the leachate from the raw sediment for the 12% mix were 14.8 and 17.6 µg/L and 2.16, 3.29 and 3.29 µg/L in the stabilized material. These results indicate that approximately 75% of the leachable arsenic is bound by the 8% Portland cement mix and approximately 80% is bound by the 12% Portland cement mix. These concentrations are greater than the Leachate Criteria for the NJ Site Specific Impact to Ground Water Soil Remediation Standards of 3 µg/L, however for the calculation of the site specific standard, dilution must be considered (NJDEP 2013). Thus with even a small amount of dilution, the site specific standard will be greater than the measured leachate concentrations indicating that that they will have no offsite impacts.

The index properties of each of the amended sediment mixes were determined prior to all chemical and physical measurements. A summary of the index properties of the amended sediments is presented in Table 2.

Table 2. Index Properties Summary

Sample	Water Content% (ASTM D2216)	LL% (ASTM D4318)	PL% (ASTM D4318)	USCS (ASTM D2487)	Organic Content (ASTM D2974)	GS (ASTM D854)
Shallow 4%	117.35	85	37	OH	8.7	2.27
Shallow 8%	80.78	72	32	OH	7.7	2.39
Shallow 12%	124.68	56	42	OH	6.2	2.53
Deep 4%	153.72	91	40	OH	8.2	2.42
Deep 8%	128.41	63	39	OH	6.9	1.94
Deep 12%	116.86	74	38	OH	6.6	2.02

The grain size distribution is presented in Figure 5.

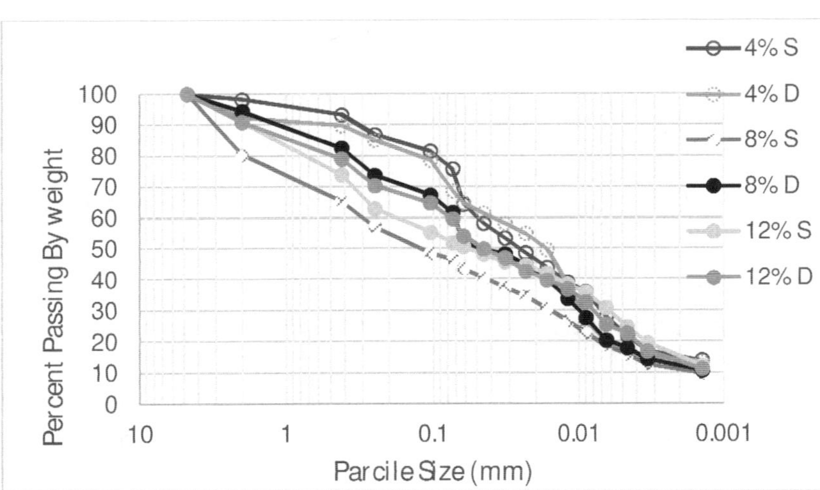

Figure 5. Grain Size Distribution

In-situ evaluation of stabilized materials were conducted using Cone Penetrometer Tests (CPT, ASTM D5778). CPT evaluation were conducted for each of the test pits to determine the vertical profile of in-situ material strength.

Figure 6. Cone Penetration Test

The results (Figure 7a.) show that the material strength is fairly uniform throughout the depth profile for each of the stabilized mixes with the 12% being the strongest. Linear regression was used to determine the relationship between tip resistance and measured UC strength for samples collected throughout the depth profile at all the test pits.

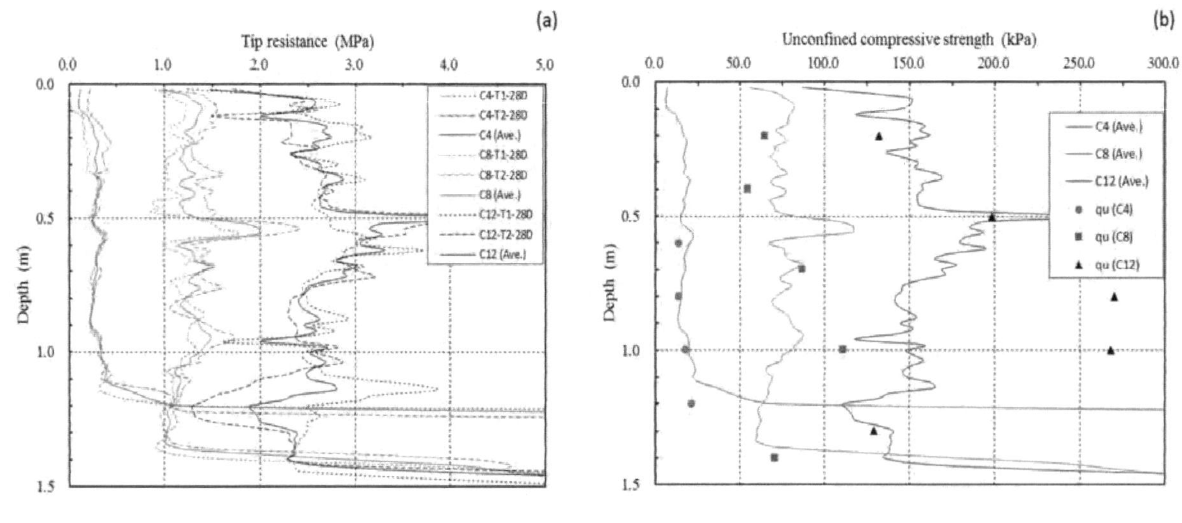

Figure 7. (a) CPT measured tip resistance throughout the stabilized dredged material profile (b) Estimated UC strength throughout the stabilized dredged material profile.

The PFTM's relatively small footprint and ability to push sediment up to 1 km inland in the processing make it an ideal mobile system for on-site S/S of impacted sediments. One potential configuration is a barge mounted system shown in Figure 8. This type of configuration would allow the treatment barge to be brought to the project site, process the sediment and transport it upland to a placement site or staging area. The benefits of this system over conventional methods is the elimination of multiple handling steps and no requirement for bulkheads or other infrastructure required for loading/unloading of scows.

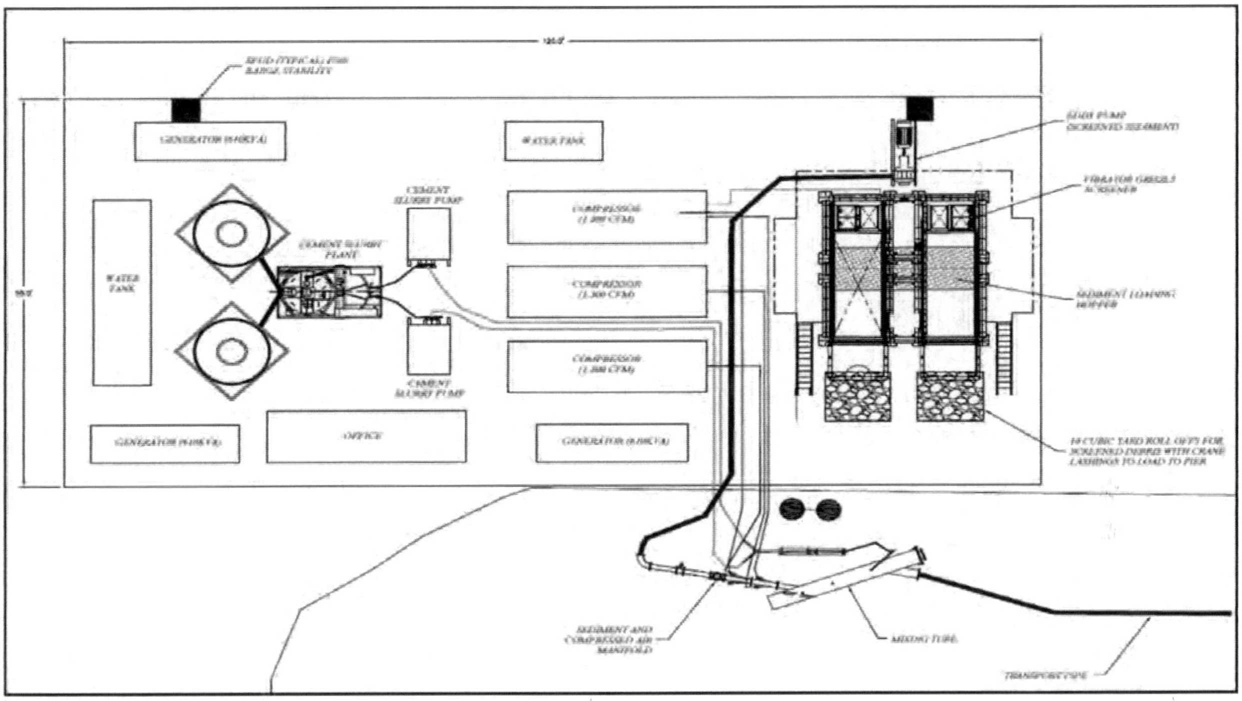

Figure 8. Barge Mounted PFTM Configuration

Conclusions

The results of this experimental program demonstrate the utility of PFTM as a rapid and efficient method for stabilization and placement of soft sediments dredged from the NY/NJ Harbor. The material stabilized using PFTM had uniform geotechnical properties and acceptable chemical properties resulting in a high potential for beneficial use. PFTM is a viable technology that can be used for the stabilization of dredged sediments. The system has attributes that may be beneficial compared with tradition processing strategies. These include the small footprint of the system, the excellent mixing provided by the method, the ability to transport material 1000 meters upland as part of the treatment process, and the enclosed/automated nature of the system which will reduce handling/exposure which may be more beneficial with more contaminated sediments.

References

Coastal Development Institute of Technology. 2008a. Technical manual of pneumatic flow mixing method, revised version, 187.

Coastal Development Institute of Technology. 2008b. Technical manual of premixing method, revised version, 215. (In Japanese)

Kitazume, M., and K. Hayano. 2007. Strength properties and variance of cement-treated ground using the pneumatic flow mixing method. Proceedings of the Institution of Civil Engineers - Ground Improvement 11:21–26. doi:10.1680/grim.2007.11.1.21

Kitazume, M., and T. Satoh. 2003. Development of a pneumatic flow mixing method and its application to Central Japan International Airport construction. Proceedings of the Institution of Civil Engineers - Ground Improvement 7:139–48. doi:10.1680/grim.7.3.139.37306

Kitazume, M., and T. Satoh. 2005. Quality control in Central Japan International Airport construction. Proceedings of the Institution of Civil Engineers - Ground Improvement 9:59–66. doi:10.1680/grim.9.2.59.63637

Ministry of Transport, The Fifth District Port Construction Bureaus. 1999. Pneumatic flow mixing method, 157. Yasuki Publishers. (In Japanese)

Mizukami, J., and Matsunaga, Y. 2015. Construction of D-Runway at Tokyo International Airport. Proceedings of the 15th Asian Regional Conference on Soil Mechanics and Geotechnical Engineering, CD.

New Jersey Department of Environmental Protection. 2013. Development of site-specific impact to ground water soil remediation standards using the Synthetic Precipitation Leaching Procedure. http://www.nj.gov/dep/srp/guidance/rs/splp_guidance.pdf (accessed February 2016).

AUTHOR INDEX

The World Association for Waterborne Transport Infrastructure (PIANC)
General Secretariat
Boulevard Roi Albert II, 20 – B. 3
B-1000 Brussels
Belgium

ISBN 978-1-5108-8140-2

CURRAN ASSOCIATES INC.
proceedings.com

International Conference on Polarised Neutrons for Condensed Matter Investigations 2014 (PNCMI 2014)

Journal of Physics: Conference Series Volume 711

Sydney, Australia
15 – 19 September 2014

Institute of Physics